Studies in Computational Intelligence

Volume 1160

The series "Studies in Computational Intelligence" (SCI) publishes new developments and advances in the various areas of computational intelligence—quickly and with a high quality. The intent is to cover the theory, applications, and design methods of computational intelligence, as embedded in the fields of engineering, computer science, physics and life sciences, as well as the methodologies behind them. The series contains monographs, lecture notes and edited volumes in computational intelligence spanning the areas of neural networks, connectionist systems, genetic algorithms, evolutionary computation, artificial intelligence, cellular automata, self-organizing systems, soft computing, fuzzy systems, and hybrid intelligent systems. Of particular value to both the contributors and the readership are the short publication timeframe and the world-wide distribution, which enable both wide and rapid dissemination of research output.

Indexed by SCOPUS, DBLP, WTI AG (Switzerland), zbMATH, SCImago.

All books published in the series are submitted for consideration in Web of Science.

Roger Lee · Gwangyoung Gim · Jongbae Kim
Editors

AI and Metaverse

Volume 2

Springer

Editors
Roger Lee
Central Michigan University
Mount Pleasant, MI, USA

Gwangyoung Gim
Soongsil University
Seoul, Korea (Republic of)

Jongbae Kim
Soongsil University
Seoul, Korea (Republic of)

ISSN 1860-949X ISSN 1860-9503 (electronic)
Studies in Computational Intelligence
ISBN 978-3-031-61162-9 ISBN 978-3-031-61163-6 (eBook)
https://doi.org/10.1007/978-3-031-61163-6

This Springer imprint is published by the registered company Springer Nature Switzerland AG
The registered company address is: Gewerbestrasse 11, 6330 Cham, Switzerland

Foreword

The main purpose of this book is to seek peer-reviewed original research papers on the foundations and new developments in Artificial Intelligence and Metaverse. The focus will also be on publishing in a timely manner, the results of applying new and emerging technologies originating from research in Artificial Intelligence and Metaverse. The finding of this book can be applied to a variety of areas, and applications can range across many fields.

The papers in this book were chosen based on review scores submitted by members of the editorial review board and underwent rigorous rounds of review.

We would like to thank all contributors including all reviewers, and all editorial board members of this book for their cooperation in helping to publish this book.

It is our sincere hope that this book provides stimulation and inspiration, and that it will be used as a foundation for works to come.

December 2023

Prof. Jongwoo Park
Soongsil University
Seoul, Korea (Republic of)

Editorial Review Board

Contents

A Study on the Influence of Supervisor's Coaching Leadership on the Innovative Behavior of Organizational Members: Focusing on the Mediating Effects of Psychological Empowerment and Job Engagement

Mi-Jeong Kim

Abstract This study aimed to verify the structural path through which a Supervisor's Coaching Leadership influences Innovative Behavior via Psychological Empowerment and Job Engagement. To this end, data were collected from employees working in domestic companies with more than 300 employees, and the research hypotheses were verified using structural equation modeling. The results showed that Coaching Leadership had a positive impact on Innovative Behavior, Psychological Empowerment, and Job Engagement. Also, Psychological Empowerment had a positive effect on Innovative Behavior and Job Engagement, and Job Engagement had a positive effect on Innovative Behavior. Furthermore, in the relationship between Coaching Leadership and Innovative Behavior, Psychological Empowerment and Job Engagement partially mediated the relationship. Implications from this study's results include that it verified the path leading to Innovative Behavior through Psychological Empowerment and Job Engagement from supervisory Coaching Leadership. It thus proposed a mechanism explaining changes in Organizational Members 'Innovative Behaviors stemming from Coaching Leadership. Through this study, it was clarified that supervisors' Coaching Leadership leads to Innovative Behaviors through Organizational Members 'Psychological changes—a motivational mechanism aspect—suggesting that corporate leaders need to be nurtured as coach-type leaders to promote Organizational Members' innovation behaviors.

Keywords Coaching leadership · Innovative behavior · Psychological empowerment · Job engagement

M.-J. Kim (✉)
Korea Smart Future Education Association, Seoul, Republic of Korea
e-mail: perfumejung@naver.com

R. Lee et al. (eds.), *AI and Metaverse*, Studies in Computational Intelligence 1160,
https://doi.org/10.1007/978-3-031-61163-6_1

1 Introduction

We live in an era of 'great change', where traditional theories and practices are becoming difficult to apply, and a people-centered management strategy has begun to be emphasized for the continuous growth of companies. With the importance of a leader's role centering on the internal factors of Organizational Members [1], the paradigm of leadership is also changing. In particular, Coaching Leadership is receiving attention as a leadership for field reflection and continuous development, emphasizing horizontal partnership differently from existing leadership [2]. However, discussions on this are relatively lacking [3]. On the other hand, management centered on Psychological factors is receiving attention, and leaders' involvement and Members 'internal factors have a decisive influence on Organizational performance. Innovative Behavior includes creative ideas from Members that greatly affect Organizational performance and their application. However, existing research lacks studies on changes in Innovative Behavior according to individual Psychological states while dealing with Innovative Behavior centered around leaders [4].

This lack of research should be addressed by focusing on individual characteristics and Psychological and cognitive differences. In previous studies, Psychological Empowerment was presented in connection with Members 'additional role expectations, which includes the process of motivating Members to have an active intrinsic motivation for their work [5]. In addition, Job Engagement has been revealed to greatly influence Organizational performance and employees' turnover intentions with its positive and achievement-oriented attitude towards work [6].

However, research on the relationship between Job Engagement and various variables that influence it is still insufficient. The purpose of this study is to explore the relationships among Coaching Leadership, Innovative Behavior, Psychological Empowerment, and Job Engagement. Specifically, it analyzes the structural path by which sub-factors of Coaching Leadership influence Innovative Behavior. Under such a background, this study aims to research the relationships among Coaching Leadership, Innovative Behavior, Psychological Empowerment and Job Engagement. Specifically aiming to analyze the structural paths related to how each variable of direction setting; competence development; performance evaluation; and relationship—all subcomponents of Coaching Leadership—affect Innovative Behavior.

Furthermore it seeks to analyze also the mediating effects of Psychological Empowerment and Job Engagement in these relationships. Through this study we aim at recognizing impacts and importance that a leader's Coaching Leadership has on forming Innovative Behaviors within companies. Furthermore we seek to provide strategic direction for qualitative improvement of Organizational Members for enhancing competitiveness as well as securing competitive advantage within companies thus providing academic as well as practical implications.

2 Theoretical Background

2.1 Coaching Leadership

Coaching Leadership was first introduced by [2] and has emerged as an important topic in contemporary leadership research. Goleman et al. [7] regarded it as important, pointing out the association between emotional intelligence and Coaching Leadership. Essentially, Coaching Leadership involves improving job competence and performance based on horizontal communication between leaders and Members, helping Members solve problems themselves. While [2] presented various approaches to the components of Coaching Leadership, they all shared a common view that Coaching Leadership has a positive impact on Organizational performance. In addition, many studies including [8] have proven various positive effects of Coaching Leadership. These research results further highlight the importance of Coaching Leadership.

2.2 Innovative Behavior

Innovation within an organization is an important activity that responds to changes in the external environment as a process of developing and implementing new ideas. Several researchers, such as [9], pointed out that the success of innovation is connected with the behavior of Organizational Members. Based on this [10], emphasized that 'Innovative Behavior' of individual Members can be driven by the entire organization. Scott et al. [9] discussed definitions and components of Innovative Behavior while [11] focused primarily on idea implementation which is one of key stages in Innovative Behavior process. In domestic research too discussions have been held regarding processes involved in Innovative Behaviors along with its importance. These studies focus on three core stages: idea generation, communication, and implementation.

2.3 Psychological Empowerment

Empowerment reflects Organizational member's Psychological state based on job execution ability and intrinsic motivation[5]. Leslie et al. [12] work emphasized employees' self-determination while other studies highlighted empowerment's influence over Organizational performance. Thomas and Velthouse [13] study suggested four factors for Psychological Empowerment which were reinterpreted by [5]: meaning, competence development, self-determination, and influence. Many studies showed Psychological Empowerment positively influenced job satisfaction and Organizational commitment; starting from [14], various researchers explored impacts from empowerment.

2.4 *Job Engagement*

Job Engagement is a concept that has received attention in positive psychology. Khan [15] interpreted it as the emotional, cognitive, and physical investment of Members when performing their jobs. Schaufeli et al. [16] classified it as job-related fulfillment, while [17] categorized it from the perspective of 'immersion' and 'interest'. Furthermore, [18] pointed out that high Job Engagement is linked to the pursuit of challenging goals and efforts to achieve them. Regarding the formation factors of Job Engagement, [15] personal, social, and work-related factors were presented as major variables. Following this were analyses through [19] JD-R model and studies on self-efficacy and optimism by [20, 21]. In Korea [22], along with [23], explored the relationship between Job Engagement and Organizational behavior as well as individual resources.

3 Research Method

3.1 *Research Model and Hypotheses*

Based on the purpose of this study, the research model and hypotheses are as follows:

In this study, the library to be analyzed were selected as follows. This study selected university characteristics of operating subjects [23] between the central object and the comparison object (Fig. 1).

3.1.1 Relationship Between Coaching Leadership, Innovative Behavior, Psychological Empowerment, and Job Engagement

Coaching Leadership, centered on a horizontal relationship with members, aims to provide directional guidance and support for capability enhancement. This has been found to significantly influence various areas such as Psychological Empowerment, Job Engagement, and innovative behavior. A series of studies have verified the effects of Coaching Leadership on the promotion of innovative behavior and improvement of job performance. In particular, the core characteristic of Coaching Leadership, capability development, appeared to be directly linked to the enhancement of Psychological Empowerment. Furthermore, the application of Coaching Leadership by leaders

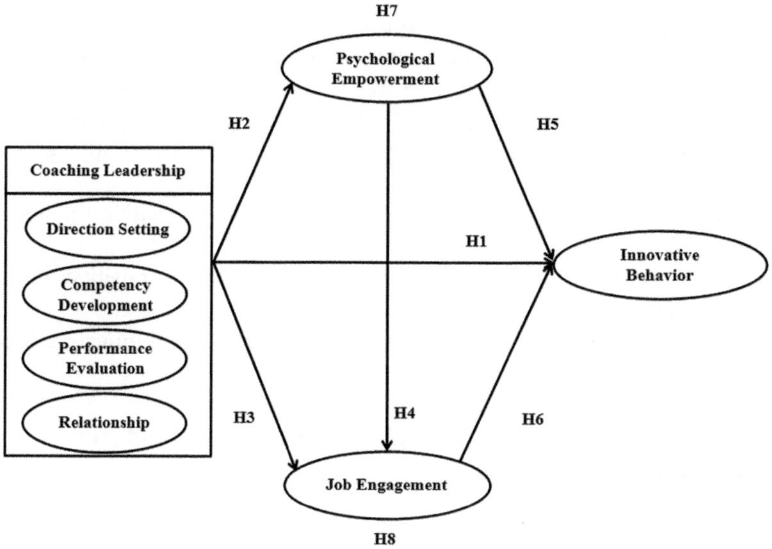

Fig. 1 Research model

with communication skills positively impacts members' Psychological Empowerment. Research has also shown that managerial coaching support plays a significant role in enhancing employees' job satisfaction and organizational commitment. Based on these studies, this study proposes hypotheses about the various effects and importance of Coaching Leadership.

Hypothesis 1: Coaching Leadership will have a positive impact on Innovative Behavior.

Hypothesis 2: Coaching Leadership will have a positive impact on Psychological Empowerment.

Hypothesis 3: Coaching Leadership will have a positive impact on Job Engagement.

3.1.2 Relationship Between Psychological Empowerment, Job Engagement, and Innovative Behavior

Psychological Empowerment refers to a mental state of confidence in job performance, and it serves as an important antecedent to Job Engagement and innovative behavior. Several studies have verified that Psychological Empowerment has a positive impact on Job Engagement. In particular, empowerment enhances employees' perceptions of autonomy and decision-making authority, and this analysis has been conducted to promote innovative behavior.

Job Engagement is a psychological features that indicates a deep state of involvement in an employee's job, and it has a positive influence on innovative behavior.

Multiple studies have confirmed that Job Engagement has a positive relationship with innovative behavior, plays an important mediating role in the relationship between supervisor support and innovative behavior, and has a significant impact on creative innovation. Based on these research findings, this study proposes new hypotheses about the importance and influence of Psychological Empowerment and Job Engagement.

Hypothesis 4: Psychological Empowerment will have a positive impact on Job Engagement.

Hypothesis 5: Psychological Empowerment will have a positive impact on Innovative Behavior.

Hypothesis 6: Job Engagement will have a positive effect on Innovative Behavior.

3.1.3 Mediating Effects of Psychological Empowerment and Job Engagement

Recent studies have shown that Coaching Leadership has a positive impact on Psychological Empowerment, and Psychological Empowerment also has a positive influence on innovative behavior. In particular, the important mediating role of Psychological Empowerment in the relationship between Coaching Leadership and innovative behavior has been emphasized. In addition, it was confirmed that coaching behavior can also influence creative behavior.

The mediating effect of Job Engagement on innovative behavior is also attracting attention in this regard. Similar leadership concepts like empowering leadership were also shown to influence innovative behavior centered on Job Engagement. There have been claims that leaders play a role in enhancing organizational commitment by appropriately managing the emotional states of organization members. These research findings have led to a better understanding of the importance of Coaching Leadership and its effects, and on this basis, new hypotheses have been proposed.

Hypothesis 7: Psychological Empowerment will mediate the influence of Coaching Leadership on Innovative Behavior.

Hypothesis 8: Job Engagement will mediate the influence of Coaching Leadership on Innovative Behavior.

3.2 Data Collection and Analysis Method

This study conducted online, face-to-face, and survey research for about two months from April 1 to May 31, in order to conduct an empirical analysis of the proposed research model targeting approximately485 employees working in companies with more than300 people nationwide. Of the returned questionnaires, we used total453 copies excluding32 missing or insincere responses for final analysis. The collected data were analyzed using SPSS24.0 and AMOS24.0 software packages. To analyze statistical models, we first conducted frequency analysis to examine demographic characteristics; secondly we performed confirmatory factor analysis for validity verification after conducting fit verification of measurement models; lastly we utilized structural equation modeling technique for hypothesis verification.

3.3 Variables and Measurement

This study deeply examined four variables—Coaching Leadership, Innovative Behavior, Psychological Empowerment, and Job Engagement. All variables were measured with seven-point Likert scale based upon operational definitions and measurement items presented in previous studies. In particular, Coaching Leadership was defined based upon [24]'s four components, and it was restructured into23 items by referencing measurement tools from various researchers. Innovative Behaviors were defined focusing upon finding new ideas, sharing them, and executing these ideas based upon concepts from [9]. Finally Job Engagements were measured with nine items based upon [16]'s research focusing upon passion and commitment towards tasks by Organizational Members.

4 Data Analysis

4.1 Characteristics of the Sample

The composition of the sample for this study is as follows: The gender distribution consists of 262 males (57.8%) and 191 females (42.2%). The age ranges from 20s to over 60s, with most being in their 30s with 164 people (36.2%) and in their 40s with149 people (32.9%). Education level varies from high school graduates or lower to bachelor's degrees, master's/doctoral degrees, with bachelor's degree holders making up the majority at343 people (75.7%). In terms of industry type, manufacturing.

4.2 Verification of the Measurement Model

In this study, a confirmatory factor analysis (CFA) was conducted to analyze the fit, reliability, and validity of the measurement model. The fit verification results for the measurement model in this study were = 1183.208 and /df = 2.491 which appeared below the recommended level of 3, RMSEA = 0.057 which appeared below the recommended level of 0.08. Also AGFI = 0.838 exceeded the recommended level of 0.8 while CFI = 0.951 and TLI = 0.945 exceeded recommended level of 0.9, indicating that the fit for our proposed measurement model was secured. Next we verified convergent validity and reliability for our measurement model. Firstly convergent validity is deemed adequate when stand ardized factor loading is above 0.5, construct reliability (CR) is above 0.7, and average variance extracted (AVE) is above 0.5 [25]. Results from convergent validity analysis in our study showed that factor loadings, CR values, and AVE values all met their respective stand ards thus securing convergent validity [26]. Reliability for a measurement model is deemed secured when internal consistency measure exceeds recommended level of 0.6 [27] with all reliabilities appearing above 0.6 indicating that they were secured (Table 1).

Indirect (mediation) effect variables in our study are Psychological Empowerment and Job Engagement. We performed additional analyses to clarify Psychological factors connecting Coaching Leadership with Innovative Behaviors along with changes in behaviors/attitudes by Organizational Members. For this we utilized

Table 1 Results from path coefficient analysis on structural model

Hyphothesis	Path			Path coefficient	Stand ard error	C. R	p
H1-1	Direction setting	→	Innovative Behavior	− 0.026	0.058	− 0.442	0.658
H1-2	Competency development	→		0.133	0.057	2.311	0.021
H1-3	Performance evaluation	→		− 0.061	0.049	− 1.239	0.215
H1-4	Relationship	→		0.137	0.057	2.388	0.017
H2-1	Direction setting	→	Psychological empowerment	0.182	0.057	3.177	0.001
H2-2	Competency development	→		0.134	0.057	2.343	0.019
H2-3	Performance evaluation	→		0.123	0.047	2.612	0.009
H2-4	Relationship	→		0.367	0.052	7.052	0.000
H3-1	Direction setting	→	Job engagement	0.131	0.054	2.436	0.015
H3-2	Competency development	→		0.113	0.053	2.137	0.033
H3-3	Performance evaluation	→		0.133	0.044	3.038	0.002
H3-4	Relationship	→		−0.065	0.053	− 1.226	0.220
H4	Psychological empowerment	→	Job engagement	0.758	0.073	10.403	0.000
H5	Psychological empowerment	→	Innovative behavior	0.392	0.138	2.832	0.005
H6	Job engagement	→	Innovative behavior	0.287	0.139	2.068	0.039

Table 2 Indirect (mediation) effects

Hyphothesis	Path					Estimate	S. E.	*p*
H7-1	Direction Setting	→	Psychological empowerment	→	Innovative behavior	0.071	0.038	0.018
H7-2	Competency development	→	Psychological empowerment	→		0.052	0.033	0.026
H7-3	Performance evaluation	→	Psychological empowerment	→		0.048	0.028	0.025
H7-4	Relationship	→	Psychological empowerment	→		0.144	0.061	0.029
H8-1	Direction setting	→	Job engagement	→	Innovative behavior	0.037	0.030	0.057
H8-2	Competency development	→	Job engagement	→		0.032	0.024	0.043
H8-3	Performance evaluation	→	Job engagement	→		0.038	0.026	0.027
H8-4	Relationship	→	Job engagement	→		− 0.019	0.023	0.153

User Defined Estimand method within AMOS using Syntax enabling succinct understanding upon individual path indirect effects, to verify hypotheses about individual path indirect effects. The results are as follows: Paths from Coaching Leadership towards Job Engagement, from Coaching Leadership towards Innovative Behaviors, and paths where individual path indirect effects where Coaching Leadership connects with Innovative Behaviors mediated by Psychological Empowerment showed that coefficients for direction guidance, capability development, performance evaluation, and relations all analyzed at $p < 0.05$ thus supporting hypotheses H7-1, H7-2, H7-3, H7-4. Paths where Coaching Leadership connects with Innovative Behaviors mediated by Job Engagement showed that only Capability Development satisfied $p < 0.05$ thus supporting hypothesis H8-1, H8-2, and H8-3. were supported, and H8-1, H8-4 were rejected (Table 2)

5 Conclusions

This study has deeply analyzed the relationship between a Supervisor's Coaching Leadership and Innovative Behavior, and explored their association by setting Psychological Empowerment and Job Engagement as mediating variables. In particular, path analysis was conducted to see what influence each sub-factor has on Innovative Behavior, and the complex relationships between variables were empirically reviewed using a structural equation model.

Firstly, Coaching Leadership has a significant impact on the growth of an organization. This is because it positively affects Members 'Psychological Empowerment, Job Engagement, and Innovative Behavior. In terms of Innovative Behavior, support from leaders and good relationships enhance the initiative and voluntariness of Members. Leaders applying Coaching Leadership value horizontal relationships and actively support independent decision-making by Members. Moreover, when leaders provide clear direction for capability development of Members along with fair feedbacks, Job Engagement also improves.

Secondly, Psychological Empowerment positively affects Job Engagement and Innovative Behaviors by enhancing confidence in their work tasks as well as proactiveness among Organizational Members. This relationship aligns with previous studies [5]. Job Engagement promotes creative innovation through role satisfaction and realization desires among Organizational Members.

Thirdly, under Coaching Leadership conditions, Psychological Empowerment and Job Engagement have mediating effects towards Innovative Behaviors. The Coaching Leadership from supervisors enhances meaning in work tasks for its employees along with confidence which boosts self-determination through authority sharing. Support from leaders along with feedback enhances task commitment along with work capabilities among its employees which promotes innovation.

In summary, the results of this study can be summarized that as clear Organizational direction from supervisors towards Organizational Members, roles, and growth, fair evaluations, and trust relations are well executed among them, Psychological Empowerment among Organizational Members is enhanced. This study suggests necessity for programs enhancing Psychological efficacy alongside fostering coach-type leaders for promoting Innovative Behaviors within organizations.

References

1. Kwang-hyun K, Donghak-rim (2019) Effect of people-centered entrepreneurship on core competencies and corporate performance of SMEs. J Korean Content Soc 19(5):217–232
2. Stowell SJ (1987) Leadership and the coaching process in organizations. Ph.D., University of Uta
3. Avolio BJ, Griffith J, Wernsing TS, Walumbwa FO (2010) What is authentic leadership development? Oxford handbook of positive psychology and work. Oxford University Press, Oxford, pp 39–51
4. Seok-bong C (2018) Employees' psychological ownership, job autonomy, and innovative behavior: a mediated moderating effect model. Corp Manage Res 25(4):65–82
5. Spreitzer GM (1995) Psychological empowerment in the workplace: dimensions, measurement, and validation. Acad Manag J 38(5):1442–1465
6. Han-young S, In J, Sun-hee O (2017) Relationship between job resources, job enthusiasm, and organizational effectiveness of employees of social enterprises. Manage Res 32(4):373–397
7. Goleman D, Boyatzis R, McKee A (2002) The emotional reality of teams. J Organ Excell 21(2):55–65

8. Joo-hyun P, Byung-seop Y (2016) The influence of Coaching Leadership on organizational commitment: the mediating effect of organizational trust. J Acad Conf Korean Assoc Professional Manage 25–38
9. Scott SG, Bruce RA (1994) Determinants of innovative behavior: A path model of individual innovation in the workplace. Acad Manag J 37(3):580–607
10. Griffin MA, Neal A, Parker SK (2007) A new model of work role performance: positive behavior in uncertain and interdependent contexts. Acad Manage J 50(2):327–347
11. Lukes M, Stephan U (2017) Measuring employee innovation: a review of existing scales and the development of the innovative behavior and innovation support inventories across cultures. Int J Entrep Behav Res 23(1):136–158
12. Leslie DR, Holzhalb CM, Holland TP (1998) Measuring staff empowerment: development of a worker empowerment scale. Res Soc Work P ractice 8(2):212–222
13. Thomas KW, Velthouse BA (1990) Cognitive elements of empowerment: an "interpretive" model of intrinsic task motivation. Acad Manag Rev 15(4):666–681
14. Goodman EA, Zammuto RF, Gifford BD (2001) The competing values framework: understanding the impact of organizational culture on the quality of work life. Organ Dev J 19(3):58–68
15. Kahn WA (1990) Psychological conditions of personal engagement and disengagement at work. Acad Manag J 33(4):692–724
16. Schaufeli WB, Salanova M, González-Romá V, Bakker AB (2002) The measurement of engagement and burnout: a two sample confirmatory factor analytic approach. J Happiness Stud 3(1):71–92
17. Rothbard NP (2001) Enriching or depleting? The dynamics of engagement in work and family roles. Adm Sci Q 46(4):655–684
18. Schaufeli WB, Bakker AB (2010) Defining and measuring work engagement: bringing clarity to the concept. Work engagement: a handbook of essential theory and research. Psychology Press, London, pp 10–12
19. Schaufeli WB, Bakker AB (2003) Utrecht work engagement scale: preliminary manual. Occup Health Psychol 162 - Unit Utrecht Univ 26(1):64–100
20. Bakker AB, Sanz-Vergel AI (2013) Weekly work Engagement and flourishing: the role of hindrance and challenge job demands. J Vocat Behav 83(3):397–409
21. Salminen S, Mäkikangas A, Feldt T (2014) Job resources and work engagement: optimism as moderator among Finnish managers. J Eur Psychol Stud 5(1):69–77
22. Jinhye C, Hyuna K (2012) Effect of job stress and personal resources of child counselors on burnout and job enthusiasm. J Korean Child Care Support Assoc 8(5):231–252
23. O'ARA (2013). An empirical study on job enthusiasm: priorities relationship between in, organizational citizenship and innovative behavior. Sogang University Business School a doctoral dissertation at an academy
24. Stowell SJ (1988) Coaching: a commitment to leadership. Traning Dev J 42(6):34–39
25. Anderson JC, Gerbing DW (1988) Structural equation modeling in practice: a review and recommended two-step approach. Psychol Bull 103(3):411
26. Bagozzi RP, Yi Y (1988) On the evaluation of structural equation models. J Acad Mark Sci 16:74–94
27. Hair JF, Black WC, Babin BJ, Anderson RE, Tatham RL (2006) SEM: confirmatory factor analysis. Multivariate Data Analysis 6:770–842

A Study on the Effect of Metaverse-Based Virtual Experience Play Education on Children's Creativity

Hoi-kyoung Bae, Ha-yeong Kim, and Tess Lee

Abstract With the development of Information and Communications Technologies (ICT), various teaching media that combine early childhood education are being developed, and in the field of early childhood education, various educational methods are being sought with a lot of interest in the virtual experience play education environment such as Augmented Reality (AR) and Virtual Reality (VR). Therefore, in this study, whether children's creativity has a positive effect on the virtual experience play education environment was confirmed through a statistical hypothesis verification procedure, and the impact of metaverse-based virtual experience play education on creativity was quantitatively summarized. The purpose of this study is to examine the educational changes of virtual experience play education and to suggest educational directions for improving the possibility and quality of use in the field of early childhood education. The study was conducted on 120 children aged 4 and 5 from four daycare centers with similar sizes in class and number of children, and the children who participated in the study are currently organizing and operating the same curriculum focusing on life topics according to the curriculum. As a research method, two-way ANOVA, multiple variance analysis (ANCOVA), and independent sample T-test were performed according to the statistical hypothesis verification procedure of the Solomon 4 group to verify the hypothesis, the moderating effect by age was analyzed, and the pure experimental effect of the Solomon 4 group was analyzed.

Keywords Virtual experience play · Creativity · Fluency · Originality · Openness · Sensitivity · Solomon 4 group

H. Bae (✉)
Korea Smart Future Education Association, Seoul, Korea
e-mail: baehk67@hanmail.net

H. Kim
MIWORKS, Seoul, Korea
e-mail: hahane07@naver.com

T. Lee
Lycée International Xavier, Seoul, Korea

R. Lee et al. (eds.), *AI and Metaverse*, Studies in Computational Intelligence 1160,
https://doi.org/10.1007/978-3-031-61163-6_2

1 Introduction

With the development of Information and Communications Technologies (ICT), various teaching media that combine early childhood education are being developed, and augmented reality (AR) and virtual reality (VR) education methods are being sought with great interest in the online education environment in the field of early childhood education. In order to prepare for the era of the 4th industrial revolution, an educational environment that can develop creativity and express and expand creative ideas is essential. It can be said that the current educational trend places importance on creativity [1]. Infants who will live in the future will live in a new environment that has never been experienced before, and changes in education to cope with these changes are needed. Research on the use of virtual reality teaching media for children, such as research on creativity of children [2] in thinking activities using computer assimilation simulations, research on the educational effects of augmented reality [3] and research using augmented reality technology for early childhood education [4] has focused on the development of smart educational contents and usability fields, including virtual reality technology. Therefore, it is necessary to consider integrated teaching methods for learning activities using various teaching media and develop learning contents and design research that induce positive effects of the learning effect. It is an important research topic to verify whether children's creativity has a positive effect on the virtual reality education environment through various experiments and observations and quantitatively reveal these verification results.

Therefore, this study attempted to verify the hypothesis by comparing the difference between infants who provided virtual experience play education through experimental observation after experiencing themselves and infants who provided play-based education, and to examine whether these virtual experience games have a positive effect on children's creativity. The specific purpose of the study is as follows. First, among the experimental design methods of metaverse-based virtual experience play education, the Solomon 4 group method is used to verify whether virtual experience play education affects children's creativity through statistical procedures. Second, we want to quantitatively reveal how much metaverse-based virtual experience play education specifically affects children's creativity through comparison between various experiments and control groups.

By rediscovering the impact of metaverse-based virtual experience play education on children from a new angle, this study aims to provide an opportunity for educational change and a strategic direction for enhancing the possibility and quality of metaverse-based virtual experience play education in the field of early childhood education.

2 Theoretical Background

2.1 *Metaverse*

There was about 30 years of development behind the evolution of the term metaverse. With the Success of the 2018 Film Ready Player One, the Concept of Metaverse has risen back to discussion. The film presents a virtual world named 'OASIS', showing many technologies such as head-mounted broadcasts (HMDs) for virtual reality (VR) rendering, ubiquitous sensing, tactile feedback, and modeling of the physical world, giving the public a sense that metaverse can be a rare opportunity [5]. The metaverse allows everyone to access the virtual world, act as their own custom avatar, and do everything they want according to the basic rules. The Metaverse Roadmap defines the metaverse as "a fusion of diverse physical reality and sustainable virtual space, diverse users to experience both while also fusing the two." [6]. The metaverse allows learners to have more experience, explore, learn, and teach, as well as work in a new world and interact with people. They can even learn or practice in practice they will experience in the real world. As discussed above, the thoughts of the metaverse will provide a new perspective on educational technology [7, 8].

2.2 *Virtual Reality*

The world's first source of "virtual reality technology" began in 1968 when Ivan Sutherland of the University of Utah studied head mounted displays (HMDs) [9]. Virtual reality is a system and related technology that builds a specific environment, situation, or virtual scenario of real life through computer modeling and helps users interact in such a virtual environment [10], and in 1938, French playwright Antonin's book "The Theater and its Double" mentioned "virtual reality space" as "a theater that makes the audience immerse in the play" [11]. From a technology perspective, virtual reality is defined as four technologies. The first is a display that imitates the virtual world through five senses, such as sight, hearing, and touch, the second is a graphic rendering system that creates a virtual field of view, the third is a tracking system that transmits the location and direction of the head and body in real time, and the fourth is a database construction system that produces and maintains a real computer model of the virtual space [12].

2.3 *Creativity*

Creativity refers to the ability to create many ideas in an original and diverse way, and creativity, a hypothetical meaning, is widely discussed by various disciplines and scholars to explain human potential. Torrance [12] defined creativity as a unique and new idea, seeing the problem from a different perspective or a new perspective [13]. Guilford [13] said that creativity is closely related to thinking, and among them, it was viewed as diffuse or divergent thinking as the basic type of thinking of creativity and defined creativity as intellectual ability [14]. Creativity is understood as a process and product of producing something unique that is new and different from the existing one, and is also considered synonymous with a high-level intellectual process or imagination [15]. In the past, creativity has tended to be viewed as an aspect of an individual's cognitive style, but in recent years, it has tended to interact with social and environmental factors, including individual personality, and expand to perspectives that include the development of creative behavioral characteristics and creative characteristics [16]. Creative characteristics can be divided into cognitive aspects that refer to originality, fluency, sophistication, and flexibility, such as thinking ability variables, and affective aspects such as openness, curiosity, sensitivity, and independence, which refer to personality variables. In addition, creativity can be said to be a universal potential of humans to solve problem situations experienced in daily life in a unique and new way.

In this study, creativity was defined as "the ability to solve problems based on new ideas as well as to come up with other opinions in a problem situation" based on the theory of foreground (2001), and it was classified into fluency, originality, openness, and sensitivity as sub-components of creativity [17].

3 Research Method

3.1 Research Hypothesis

Torrance [17] has advocated for the purpose of early childhood education since the beginning and said that the purpose of early childhood education is to help children develop their potential and perform their social functions skills [18]. Downey et al [19] explains that while learning creative problem, the quality and quantity of ideas, originality and fluency can be considered at the same time. Therefore, the need for educational programs or teaching methods that can develop children's creativity has been highlighted. Since the 1970s, research has been conducted on the development and effectiveness of various techniques and educational programs to promote children's creativity. The results of the research usually report the positive effects of creativity programs, suggesting the possibility of creativity education.

Assuming that virtual experience play education will have a significant impact on children's creativity development, the following hypothesis is established and confirmed through experiments.

H1. Virtual experience play education will have a positive (+) effect on children's creativity.

H1.1. Virtual experience play education will have a positive (+) effect on children's fluency.

H1.2. Virtual experience play education will have a positive (+) effect on children's originality.

H1.3. Virtual experience play education will have a positive (+) effect on children's openness.

H1.4. Virtual experience play education will have a positive (+) effect on children's sensitivity.

3.2 Research Design

3.2.1 Solomon Four-Group Experimental Design

Forty years ago Solomon introduced a new type of experimental design, commonly called the Solomon Four Group design today [20]. Gung-geun [20] stated that Solomon Group 4 experimental design is the type of experimental design that provides the strongest means of control for factors that hinder internal and external validity, such as the effect of extrinsic variables, the effect of pre-inspection, and the interaction effect of inspection and treatment [21]. The design of the experimental study is shown in Fig. 1.

Statistical hypothesis verification was performed by applying a two-way ANOVA (Two-way ANOVA) considering whether the experiment was conducted and whether the pre-investigation was conducted according to the method suggested by Braver and Braver [21]. For verification, multiple variance analysis verification can be performed to conclude that the experiment has some effect if the significance verification result is significant, and if not, T verification can be performed. If the significance test results in the verification are significant, it can be concluded that the experiment is effective, and if not, there is no evidence of the experimental effect.

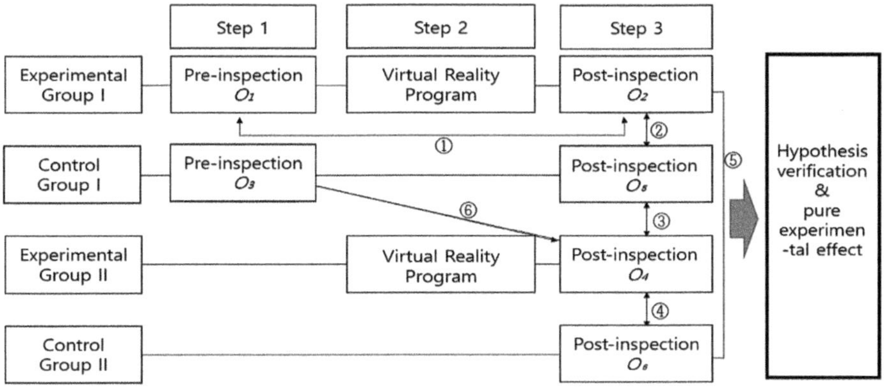

Fig. 1 Solomon 4 group analysis chart

3.2.2 Subject of Study

This study selected four daycare centers in Gyeonggi-do and conducted them on 120 children aged 4 and 5 years old. The size of the four randomly extracted daycare centers is similar to the number of classes and children, and the children who participated in the study currently organize and the same curriculum based on the life topic according to the Nuri process, and it consists of those who fully explained the purpose of this study to the guardians of the four daycare centers and submitted the consent of the guardians. The background of the experimental group was divided into experimental groups I and II, which conduct virtual experience play by installing a virtual experience play educational environment in daycare centers, and control groups I and II, which conduct daily activities under the life theme of the general Nuri curriculum.

3.3 Research Procedure

This study was conducted over a total of 19 weeks in the order of literature review, preliminary research, inspector training, teacher training, pre-test, main study, and post-test, and the research procedure is shown in Table 2.

Table 1 Gender and age of infants

Sortation	N	Gender		Age	
		Male	Female	4 years old	5 years old
Experimental group I	30	17	13	15	15
Experimental group II	30	11	19	15	15
Control group I	30	16	14	15	15
Control group II	30	14	16	15	15

Age: 4 Age: 1 January to 31 December 2017 / 5 Age: 1 January to 31 December 2016

Table 2 Research procedures

Research procedure		Schedule	Content
Review of the literature			• Literature review and selection of research tools – Measuring tools: children's creativity
Preliminary research		April 4th~April 29th, 2022	• Validation of the suitability of measuring tools – Subjects: 10 5-year-old children in E daycare center, 10 5-year-old children in S daycare center who are not test subjects – Contents: Creativity test tools and experimental groups to examine the appropriateness of experimental treatment and virtual experience play period
Inspector training		May 1, 2022	• Conduct infant creativity evaluator training – Subjects: Master's degree in Early Childhood Education, 2 people currently in doctoral program
Evaluator training		May 2nd to May 4th, 2022	• Conduct child evaluator training – Subject: 8 homeroom teachers of the experimental institution
Main study	Pre-inspection	May 9th to May 12th, 2022	• Conducting a pre-inspection – 15 4-year-old infants and 15 5-year-old infants in experimental group I. and control group I – Contents: Creativity pre-test
	Experimental treatment	May 16 to July 22, 2022 (10 weeks)	• Experimental group: A program conducted by an early childhood education institution and a metaverse-based virtual experience play program twice a week • Control group: Programs conducted by early childhood education institutions
	Post-inspection	July 25th to July 27th, 2022	• Conduct a follow-up inspection – 15 4-year-old infants and 15 5-year-old children in experimental group I., II. Control group I., II – Content: Creativity post-test
Data analysis and discussion			• Data analysis on creativity of virtual experience play • Discussion based on research findings

4 Research Result

In the process of analyzing the experimental design, Braver and Braver [21] presented the statistical processing process of the four Solomon groups. The Solomon 4-group experimental design verified the research hypothesis by conducting two-way ANOVA using the post-investigation score. Therefore, this study also verified the research hypothesis of creativity and playability according to this method.

Table 3 Summary of Experimental Study

	Path	Hypothesis	Adoption status
Inventiveness	H1.1	Virtual experience play education will affect children's fluency	Adoption
	H1.2	Virtual experience play education will affect children's originality	Adoption
	H1.3	Virtual experience play education will affect children's openness	Adoption
	H1.4	Virtual experience play education will affect children's sensitivity	Rejection

4.1 Verifying the Creativity Hypothesis

This study derived hypotheses to determine whether the sub-factors of children's creativity affect fluency, originality, openness, and sensitivity in virtual experience play education, and used the statistical hypothesis verification procedure of the four Solomon groups to statistically verify this. The hypothesis that virtual experience play education will have a positive (+) effect on children's fluency, original writing, and openness was conducted with ANOVA and T-test according to the analysis procedures presented in Chap. 3 to confirm the effect of pre-test and virtual experience play education. The path to hypothesis testing is shown in Fig. 2.

Therefore, it can be concluded that virtual experience play education has a statistically significant effect on increasing fluency, originality and openness and hypotheses H1.1, H1.2, and H1.3 were adopted. The hypothesis that virtual experience play

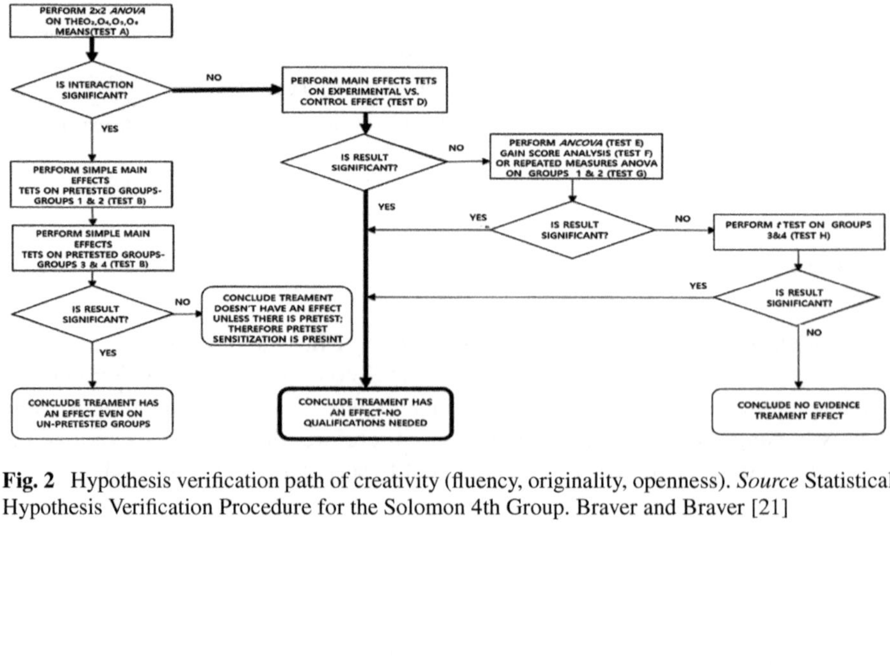

Fig. 2 Hypothesis verification path of creativity (fluency, originality, openness). *Source* Statistical Hypothesis Verification Procedure for the Solomon 4th Group. Braver and Braver [21]

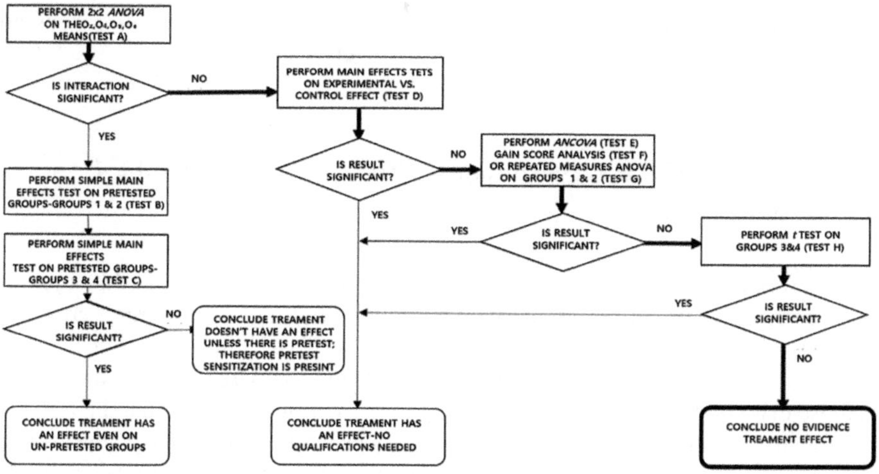

Fig. 3 Hypothesis verification path of creativity (sensitivity). *Source* Statistical hypothesis verification procedure for the Solomon 4th group. Braver and Braver [21]

education will have a positive (+) effect on children's sensitivity is that ANOVA and T-test were conducted according to the analysis procedures presented in Chap. 3 to confirm the effect of pre-test and the effect of virtual experience play education. The path to pothesis testing is shown in Fig. 3.

In order to confirm whether the virtual experience play educational activities are effective in sensitivity, it was confirmed whether the interaction between experiment status and prior status was significant. The experimental-pre-irradiation interaction was analyzed to be statistically insignificant with $F = 1.418$ and $p = 0.236$. In the analysis stage so far, the effects of the preliminary investigation were not reflected, so to accurately confirm the effect of sensitivity, an independent sample T-test of experimental group I and control group I was conducted to compare the pre-test and post-test by reflecting the pre-test scores. As a result of the analysis, the pre-test and post-test of the experimental group I and the control group I showed no significant difference on average with $t = 1.963$ and $p = 0.052$.

The comparison between Experimental Group I and Control Group I does not conclude that virtual experience play education did not affect sensitivity because the pre-test and post-test of Experimental Group I and Control Group I were not significant.

Therefore, as a result of performing covariance analysis (ANCOVA) for the following experimental group I and control group I, there was no statistically significant difference in the experimental effect with $F = 2.507$ and $p = 0.116$. As shown in Fig. 3, T-test was performed on experimental group II and control group II, and the comparison of the post-investigation scores of experimental group II and control group II, which had not been pre-tested, showed that the average difference between the two groups was not significant with $t = 0.283$ and $p = 0.779$. The results of this analysis process can be concluded that virtual experience play education does not affect sensitivity, and hypothesis H1.4 was rejected.

4.2 Analysis of the Experimental Effects of the Solomon 4 Group Design

The data collected using the SPSS 22.0 program was analyzed to verify the impact of metaverse-based virtual experience play education programs on children's creativity and playability. In the process of analyzing these four Solomon groups, it was confirmed whether the average of groups that were compared to each other, such as those that conducted virtual experience play education and those that did not, was statistically significant. For this, statistical methods were used to analyze the average difference between groups. As an analysis method to confirm the difference in the mean between these groups, a two-group independent sample T-Test analysis was conducted. Pure experimental effect values indicate that fluency, originality, openness, and sensitivity items among virtual experience play education have a positive effect on creativity through program implementation when the score is positive. The average value for creativity is 50 points, and the pure experimental effect value is 14.033, originality 6.670, openness 15.850, sensitivity 0.970, and total score 4.457, indicating that creativity has improved since the implementation of the program. As a result of quantitatively examining the pure experimental effect, openness was the highest in creativity, followed by fluency and originality, and although sensitivity was insufficient, the experimental effect was found.

5 Conclusion

In this study, as a result of examining the effects of virtual experience play education on fluency, originality, sensitivity, and openness among the sub-areas of creativity, the average scores of fluency, originality, and openness improved and had a significant effect compared to the control group. It seems to be the result of the formation of an open attitude to cooperate with peers to solve problems during virtual experience play, seek problem-solving solutions in various ways, and expand play. During the virtual experience play education, infants showed a desire, positive thoughts, and activism to experience various experiences about given problems and situations. The results of this study in the attitude of infants are in line with the findings [22] that a given environment promotes the expression of creativity so that the openness to challenge and try new experiences and solutions can be found in various situations. The results of sensitivity, another sub-area of creativity, showed that the experimental group that conducted virtual experience play was not statistically significant compared to the control group that conducted general Nuri curriculum educational activities, so this experiment did not show any effect on improving infant sensitivity.In the virtual experience play education conducted in this study, rather than focusing on virtual reality play on a large screen, observing surrounding objects, discovering problems, and sensitively observing the other person's feelings and attitudes in new exploration activities, sensitivity has not improved. Putting the above results together, it can be seen that virtual experience play education is effective for fluency and openness, which are sub-areas of creativity, but not for sensitivity.

Based on the results of this study, the academic implications are as follows.

First, virtual experience play education was conducted using the Solomon 4-group method among experimental design methods, and statistical procedures demonstrated that virtual experience play had a positive effect on fluency, originality, and openness, which are sub-factors of creativity, but not on sensitivity. Second, the effect of virtual experience play education on creativity was quantitatively revealed through comparison between the experimental group and the control group. Creativity was found to have an effect in the order of openness, fluency, originality, and sensitivity.

The significance of this study is that the use of advanced digital technology has a positive impact on creativity in the early childhood curriculum, and by rediscovering the impact of virtual experience play education on children from a new angle, it is necessary to further expand the play environment of children to a new educational transition and a more diverse and systematic national curriculum.

References

1. Kim KY (2018) Creative thinking and innovative business modeling (needed in the era of the fourth industrial revolution)
2. Kim OM (2001) Effects of thinking activities using computer assimilation simulations on infants' creativity. Doctoral dissertation. Dong-A University Graduate School. Lee J, Bagheri B, Kao HA (2015) A cyber-physical systems architecture for industry 4.0-based manufacturing systems. Manufacturing Lett 3:18–23
3. Limniou M, Smith M (2010) Teachers' and students' perspectives on teaching and learning through virtual learning environments. Eur J Eng Educ 35(6):645–653
4. Spielberg S, Silvestri A, Penn Z, Cline E, De Line D (2018) Ready player one. Warner Bros, USA
5. ASF (2006) Metaverse roadmap overview. http://www.metaverseroadmap.org/overwiew/
6. Díaz J, Saldaña C, Avila C (2020) Virtual world as a resource for hybrid education. Int J Emerg Technol Learn (iJET) 15(15):94–109
7. Rospigliosi PA (2022) Metaverse or Simulacra? Roblox, Minecraft, meta and the turn to virtual reality for education, socialisation and work. Interact Learn Environ 30(1):1–3
8. Sutherland IE (1968) A head-mounted three dimensional display. In: Proceedings of the December 9–11, 1968, lescent psychiatry. Blackwell, London, pp 285–307
9. Zeltzer D (1992) Autonomy, interaction, and presence. Presence Eleoperators Virtual Environ 1(1):127–132
10. Artaud A, Richards MC (1958) The theater and its double... translated... by Mary Caroline Richards, 5th printing. New York
11. Brooks JG, Brooks MG (1999) In search of understanding: the case for constructiveness classrooms. Ascend
12. Torrance EP (1962) Guiding creative talent. Prentice Hall Inc., NY

13. Guilford JP (1950) Creativity. Am Psychol 5(3):444–454
14. Kyung-hee J (2007) Creative learning activities for infants. Taeyoung Publishing House
15. Mellou E (1996) Can creativity be nurtured in young children? Early Child Dev Care 119(1):119–130
16. Kyung-won J (2001) Expert guide to infant drawing creativity testing. School Governor, Seoul
17. Torrance EP (1979) The search for Satori and creativity. Creative Education Foundation, NY
18. Parnes SJ (1975) Education and creativity. In: Vernon PE (ed) Creativity. Penguin, NY, pp 341–354
19. Downey G, Freitas AL, Michaelis B, Khouri H (1998) The self-fulfilling prophecy in close relationships: rejection sensitivity and rejection by romantic partners. J Pers Soc Psychol 75(2):545–560
20. Geun Geun N (2015) Administrative investigative methodology. A Law Enforcement Official, Seoul
21. Braver MW, Braver SL (1988) Statistical treatment of the Solomon four-group design: a meta-analytic approach. Psychol Bull 104(1):150–154
22. Lyubomirsky S, King L, Diener E (2005) The benefits of frequent positive affect: does happiness lead to success?. Psychol Bull, 131(6): 803
23. Solomon RL (1949) An extension of control group design. Psychol Bull 46(2):137–150

A Study on the Effects of Perceived Value of Live Commerce and Presence on Customer Satisfaction and Purchase Intention

Mi-Sook Han, Dong-Hyuk Jo, Sung-Hee Baek, and Ha-Yeong Kim

Abstract In an era where the boundaries between reality and the virtual world collapse, a new form of distribution industry is evolving. Companies have adapted their distribution structures in line with the trends of the time. The advancement in information technology has enabled smartphones to resolve almost everything, facilitating consumers to watch video content like YouTube and live streaming anywhere and engage in shopping. This shift has led to the rise of live commerce, a convergence of real-time interaction and commerce. The growth rate of this sector is remarkable. The objective of this study is to investigate how the perceived value and sense of presence in live commerce influence purchase intentions through customer satisfaction in online shopping. The analysis revealed several significant findings. First, within perceived value in live commerce, hedonic value and the sense of presence significantly affect customer satisfaction, while utilitarian value does not. Second, perceived value and the sense of presence significantly influence purchase intentions. Third, customer satisfaction significantly impacts purchase intentions. Fourth, customer satisfaction positively mediates the relationship between perceived value, sense of presence, and purchase intentions. This research provides valuable insights for companies seeking to maximize profits by creating distinctive customer value through the establishment of optimized environments and services for live commerce users.

M.-S. Han · S.-H. Baek
Department of Program in Project Management, Soong Sil University, Seoul, South Korea
e-mail: sookhan01@naver.com

S.-H. Baek
e-mail: b0s0h073@naver.com

D.-H. Jo (✉)
Department of Business Administration, Soongsil University, Seoul, Korea
e-mail: joe@ssu.ac.kr

H.-Y. Kim
MIWORKS, Seoul, South Korea
e-mail: hahane07@naver.com

R. Lee et al. (eds.), *AI and Metaverse*, Studies in Computational Intelligence 1160,
https://doi.org/10.1007/978-3-031-61163-6_3

Keywords Live commerce · Perceived value · Utilitarian value · Hedonic value · Presence · Customer satisfaction · Purchase intention

1 Introduction

As the real world and the virtual world converge, a new form of distribution industry is appearing. Consumers increasingly favor cheaper and superior products, more convenient purchasing channels, and swifter deliveries. To meet these consumer demands, businesses have adapted their distribution structures to align with the zeitgeist. The advancement in information and communication technologies has made smartphones capable of resolving virtually everything, resulting in smartphones becoming a significant mass medium in the twenty-first century. Additionally, high-speed internet has spurred the activation of video content, leading to an increase in consumers using smartphones to watch video content, such as YouTube and live broadcasts, and to engage in shopping. Recently, due to the social distancing policies brought about by COVID-19, non-face-to-face mobile e-commerce utilizing video content has rapidly expanded. This has brought attention to live commerce, a convergence of real-time interaction and trade, as a new form of e-commerce, and its growth trajectory is astonishing.

Live commerce, which originated in China, has reshaped the retail distribution structure in the Chinese market over the last few years [1]. The domestic live commerce market has experienced substantial growth since its inception in 2020. The market, which was worth 400 billion KRW in 2020, grew to an estimated 2.8 trillion KRW in 2021, marking a 700% growth, and is projected to reach 10 trillion KRW in 2023, a growth of 387% from 2021. With the current growth rate, it is forecasted to expand to 25 trillion KRW by 2025, a 250% growth from 2023 [2]. Despite Korean companies acknowledging the growth and value of the Chinese market and fiercely competing to adopt and lead it, academia lacks extensive research in this area. A more comprehensive perspective is required to understand the variables that drive live commerce users' purchasing behavior. As live commerce transforms e-commerce from a product-centered environment to a social, pleasure, and customer-centric environment [3], the perceived shopping value of live commerce is recognized as a strong force in the market for understanding consumer behavior [4]. Perceived value is presented as the functional utility of shopping (utilitarian value) and the experiential benefit of shopping (hedonic value).

The purpose of this study is to investigate how perceived value from the consumer's perspective, along with the sense of presence in online shopping, influences purchase intentions through customer satisfaction among live commerce users. To achieve this objective, this study aims to develop a new research model that focuses on perceived value and the sense of presence in live commerce and their impact on purchase intentions, while validating the mediating effect of customer satisfaction. By empirically analyzing this model, this study seeks to differentiate from previous research and provide insights valuable for businesses aiming to maximize profits by establishing an optimized environment and services tailored for live commerce users. Furthermore, the rapidly changing distribution channels have altered consumers' information search processes. Thus, the need to explore various environmental factors influencing information search has emerged. By presenting the unique characteristics of live commerce in contrast to traditional online commerce, this research also aims to offer practical information about consumers' information processing to adapt to the changing information search process.

2 Theoretical Background

2.1 Perceived Value of Live Commerce

Live commerce is a portmanteau of live streaming and e-commerce, representing a mobile shopping platform where real-time sales and purchases take place through online live broadcasts [5].

Through shopping can be considered an overall evaluation of multidimensional shopping experiences [6]. Early research on shopping value by Babin, Darden, and Griffin segmented shopping value into utilitarian value and hedonic value [7].

Utilitarian value refers to the value related to functional benefits in shopping experiences, such as monetary savings, time savings, and convenience. The perspective explaining utilitarian value emphasizes consumers as rational problem solvers, highlighting functional and goal-oriented thinking. From this standpoint, shopping can be seen as a means to fulfill intended purposes. Thus, criteria such as monetary savings or convenience serve as instrumental and extrinsic factors in value assessment [7].

Hedonic value, on the other hand, signifies the value provided by experiential benefits, such as fun and enjoyment, experienced during the shopping process. Hedonic value involves abstract and subjective value assessments because, from the viewpoint that shopping itself is the goal, not just a means to fulfill consumer purposes, the evaluation is based on the enjoyment, pleasure, escapism, and exploration that can be obtained through shopping [7]. Hence, based on previous research, the following hypotheses were formulated.

Hypothesis 1. Perceived value of live commerce will have a positive (+) impact on customer satisfaction.

Hypothesis 1.1. Utilitarian value will have a positive (+) impact on customer satisfaction.

Hypothesis 1.2. Hedonic value will have a positive (+) impact on customer satisfaction.

Hypothesis 2. Perceived value of live commerce will have a positive (+) impact on purchase intention.

Hypothesis 2.1. Utilitarian value will have a positive (+) impact on purchase intention.

Hypothesis 2.2. Hedonic value will have a positive (+) impact on purchase intention.

2.2 Presence

Presence is defined as the perception of direct experiences that consumers undergo through technology in a communication environment [8]. Presence appears to be experienced even in virtual spaces, such as in live commerce or when enjoying content or using online shopping malls [9]. Ou et al. redefined presence in the context of online commerce, emphasizing the formation of simultaneity and intimacy between consumers and sellers [10]. The sense of presence in shopping situations is a subjective experience for consumers and is perceived in the virtual space mediating online media, influenced by the similarity to the actual offline shopping environment [11]. Based on prior research, the following hypotheses are formulated.

Hypothesis 3. Presence will have a positive (+) impact on customer satisfaction.

Hypothesis 4. Presence will have a positive (+) impact on purchase intention.

2.3 Customer Satisfaction

Huff defined customer satisfaction as the supplier exceeding customer expectations, and customers being satisfied beyond their expectations [12]. Oliver defined customer satisfaction as the customer's response and judgment regarding how well their demands have been met [13]. From a business perspective, customer satisfaction is one of the factors that must be maintained to enhance sales and increase profits. Meeting customer satisfaction can lead to increased profits for the company and secure competitiveness by enhancing the intention of repeat purchases. Customer satisfaction is recognized as an important factor in the consumer behavior process, inducing favorable emotions during product and service purchases and influencing the intention to repurchase [14].

In this study, customer satisfaction is defined as continuously maintaining and strengthening trust, preference, and affinity for products and services by initially meeting or exceeding customers' demands or expectations [15]. Based on previous research, the following hypotheses are formulated.

Hypothesis 5. Customer satisfaction will have a positive (+) effect on purchase intention.

Hypothesis 6. Customer satisfaction will positively mediate the relationship between perceived value of live commerce and purchase intention.

Hypothesis 6.1. Customer satisfaction will positively mediate the relationship between utilitarian value and purchase intention.

Hypothesis 6.2. Customer satisfaction will positively mediate the relationship between hedonic value and purchase intention.

Hypothesis 7. Customer satisfaction will positively mediate the relationship between telepresence and purchase intention.

2.4 Purchase Intention

The ultimate purpose of live commerce is the creation of exchanges, namely, eliciting purchasing behavior. Information conveyed by the message sender in advertising or personal selling can change the recipient's attitudes toward a specific product, resulting in the occurrence of purchase intentions and purchase behavior. Purchase intention represents the consumer's planned behavior, indicating the willingness to move from intention to actual purchase behavior. It reflects the phenomenon where a consumer's perceived values, beliefs, attitudes, etc., toward a specific target are manifested in behavior [16]. The likelihood of purchase behavior increases when consumers perceive higher values, as confirmed by numerous marketing literature [17]. Perceived value plays a role as a precursor that can predict future behavioral intentions because it acts as a powerful motivator [18]. In the process of efficiently achieving the intended purchase goals through live commerce, if consumers perceive utilitarian shopping value or experience high hedonic value, such as enjoyment, fun, and escapism during the shopping process, the intention to purchase products through live commerce is likely to increase [19].

3 Research Method

3.1 Research Model

Based on a review of relevant prior studies, this study constructs the independent variable, perceived value of live commerce, with two sub-concepts: utilitarian value and hedonic value, and includes a further component, presence. Customer satisfaction is constructed as a mediating variable, and purchase intention is set as the dependent variable. The conceptual model of this study, assuming that the perceived value and presence of live commerce influence purchase intention through the mediation of customer satisfaction, is illustrated in Fig. 1.

3.2 Operational Definitions and Measurement Items

Based on prior research, operational definitions for variables were established, and corresponding items were adopted to construct the survey. The survey items were

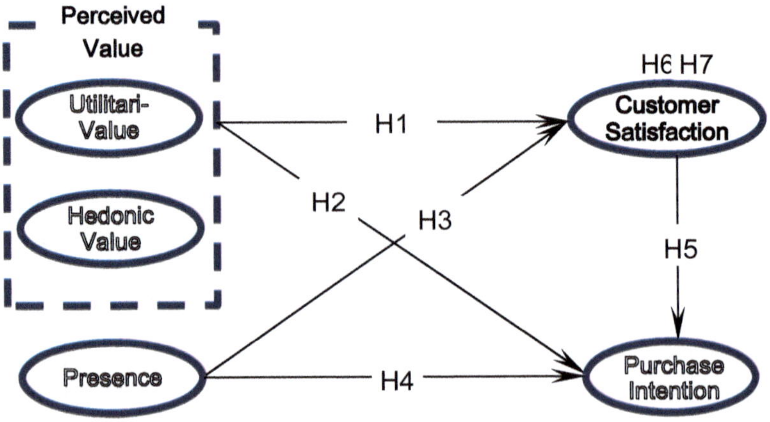

Fig. 1 Research model

structured using a 7-point Likert scale, ranging from '1—strongly disagree' to '7—strongly agree (Table 1).'

Table 1 Operational definitions

Variable		Operational definition	References
Perceived value	Utilitarian value	Value related to functional benefits of shopping experiences during live commerce use, such as financial savings, time savings, and convenience	[7]
	Hedonic value	Value derived from experiential benefits, such as fun and enjoyment, during live commerce use	[7]
Presence		Perception of direct experiences consumers have through technology in virtual spaces like live commerce	[8]
Customer satisfaction		Condition induced by fulfilling customers' original demands or surpassing their expectations, encouraging purchase, repurchase, and repeated buying behavior	[15]
Purchase intention		Consumer intentions reflecting the values, beliefs, and attitudes perceived for a specific target during live commerce use	[16]

4 Analysis Result

This study aimed to empirically investigate the impact of perceived value and presence in live commerce on customer satisfaction and purchase intention. A survey was conducted targeting customers with experience in live commerce. Data collection took place from May to June 2023, spanning approximately two months, through on-site visits and online messenger services. A total of 220 responses were collected, and after excluding cases with missing data or unreliable responses (7 cases), the final sample comprised 213 valid responses for analysis.

Examining the sample characteristics of this study, the gender distribution was 105 males (49.3%) and 108 females (50.7%). Regarding age groups, there were 38 individuals in their 20s (17.8%), 40 in their 30s (18.8%), 58 in their 40s (31.9%), 58 in their 50s (27.2%), and 9 aged 60 and above (4.2%). The duration of live commerce usage revealed that 38.0% had used it for less than 6 months, 21.1% for 6–12 months, 11.7% for 12–18 months, 4.2% for 18–24 months, and 24.9% for 24 months or more. Regarding the frequency of live commerce usage, 58.2% made purchases less than once a month, 27.7% 2–3 times a month, 5.6% once a week, and 8.5% more than 2–3 times a week. The majority of participants (58.2%) reported a single purchase cost of 50,000 to 100,000 won.

4.1 Reliability and Validity Analysis of Measurement Tools

In this study, exploratory factor analysis was conducted to verify the reliability and validity of the measurement tools, and internal consistency was confirmed using Cronbach's alpha coefficient. The validation results indicated that the factor loading of measurement items was all above 0.5, showing an appropriate level. Additionally, Cronbach's alpha coefficients were all above 0.7, indicating overall reliability and validity of the measurement tools were secured (Table 2).

4.2 Correlation Analysis

To examine the relationships between variables in the proposed model, Pearson's correlation analysis was conducted. The results, as shown in Table 3, indicate positive (+) correlations between all variables. This observation aligns with the assumptions made for hypothesis testing in the present research model.

Table 2 Reliability and validity analysis

Variable		Factor loading					Cronbach's α
		Factor 1	Factor 2	Factor 3	Factor 4	Factor 5	
Purchase Intention	Purchase intention 4	**0.792**	0.315	0.166	0.077	0.087	**0.871**
	Purchase intention 3	**0.783**	0.250	0.189	0.134	0.185	
	Purchase intention 2	**0.754**	0.136	0.118	0.283	0.259	
	Purchase intention 1	**0.712**	0.060	0.086	0.260	0.335	
Hedonic Value	Hedonic value 1	0.130	**0.734**	0.265	0.137	0.107	**0.812**
	Hedonic value 2	0.265	**0.733**	0.147	0.196	0.127	
	Hedonic value 3	0.130	**0.680**	0.145	0.226	0.163	
	Hedonic value 4	0.241	**0.588**	0.184	0.330	0.260	
Presence	Presence 4	0.143	0.048	**0.801**	0.138	0.121	**0.823**
	Presence 3	0.030	0.159	**0.775**	0.156	0.247	
	Presence 2	0.164	0.352	**0.718**	0.067	0.033	
	Presence 1	0.148	0.169	**0.712**	0.024	0.220	
Utilitarian value	Utilitarian value 2	0.121	0.102	0.019	**0.761**	0.114	**0.721**
	Utilitarian value 4	0.258	0.045	0.058	**0.682**	0.138	
	Utilitarian value 3	0.077	0.319	0.140	**0.647**	− 0.060	
	Utilitarian value 1	0.105	0.276	0.167	**0.639**	0.086	
Customer Satisfaction	Customer satisfaction 2	0.201	0.091	0.294	0.074	**0.755**	**0.776**
	Customer satisfaction 1	0.215	0.021	0.250	0.046	**0.706**	
	Customer satisfaction 3	0.134	0.369	0.070	0.224	**0.681**	
	Customer satisfaction 4	0.313	0.385	0.066	0.038	**0.557**	
Eigen value		2.848	2.732	2.707	2.367	2.353	
Variance explanation (%)		14.239	13.661	13.534	11.836	11.765	

(continued)

Table 2 (continued)

Variable	Factor loading					Cronbach's α
	Factor 1	Factor 2	Factor 3	Factor 4	Factor 5	
Kaiser–Meyer–Olkin measure	0.896					
Bartlett's test of sphericity (p)	0.000					

The values in bold are the Factor loading values for the Factors (Factors 1–5)

Table 3 Correlation analysis

Variable	M	SD	1	2	3	4	5
1. Utilitarian value	5.0998	0.83112	1				
2. Hedonic value	4.7359	0.93531	0.544**	1			
3. Presence	4.7019	0.96841	0.333**	0.504**	1		
4. Customer satisfaction	4.9883	0.83368	0.363**	0.541**	0.492**	1	
5. Purchase intention	4.8498	0.98231	0.470**	0.555**	0.412**	0.591**	1

** $p < 0.01$

4.3 Hypothesis Testing Results

The hypothesis testing results for the impact relationships between perceived value, presence, customer satisfaction, and purchase intention in live commerce are presented in Table 4.

To investigate the mediating effects of customer satisfaction in the relationship between perceived value, presence, and purchase intention in live shopping services, a mediation analysis was conducted. The results, presented in Table 5, indicate that hedonic value partially mediates customer satisfaction, while presence fully mediates customer satisfaction, significantly influencing purchase intention. Consequently, Hypotheses 6 and 7 are supported.

Table 4 Hypothesis testing results

Dependent variable	Independent variable	Unstandardized coefficient		Standardized coefficient	T-value	P-value	Result
		β	Standard error	β			
Customer satisfaction	Utilitarian value	0.073	0.066	0.073	1.101	0.272	Not supported
	Hedonic value	0.318	0.064	0.357	4.937**	0.000	Supported
	Presence	0.247	0.055	0.287	4.473**	0.000	Supported
Purchase intention	Utilitarian value	0.267	0.078	0.226	3.430**	0.001	Supported
	Hedonic value	0.369	0.075	0.352	4.893***	0.000	Supported
	Presence	0.162	0.065	0.160	2.500*	0.013	Supported
Purchase intention	Customer satisfaction	0.697	0.065	0.591	10.655	0.000	Supported

*$p < 0.05$**$p < 0.01$, ***$p < 0.001$, ( ) denotes standardized coefficient (β)

Table 5 Verification of mediation effects of customer satisfaction

Category	Step 1	Step 2	Step 3
	Customer satisfaction	Purchase intention	Purchase intention
Utilitarian value	1.101(0.973)	3.430(0.226)**	3.238(0.198)*
Hedonic value	4.937(0.357)**	4.893(0.352)**	3.085(0.217)*
Presence	4.473(0.287)**	2.500(0.160)*	0.831(0.052)
Customer satisfaction			5.899(0.377)**

*$p < 0.05$, **$p < 0.01$, ( ) denotes standardized coefficient (β)

5 Conclusion

This study aimed to verify the impact of the perceived value and sense of reality of live commerce on customer satisfaction, which, in turn, influences purchase intentions among consumers with experience in live commerce, a topic gaining attention in recent marketing communication. Based on previous research, research hypotheses and a research model were proposed and empirically analyzed.

The results of this study can be summarized as follows. First, among the perceived values of live commerce, hedonic value and sense of reality were found to have a significant impact on customer satisfaction, while utilitarian value did not show a significant influence. This aligns with previous research results indicating that utilitarian value leads to trust in the seller and electronic word-of-mouth (E-WOM) when obtaining practical and realistic current values through easy and accurate access to product information [20]. Second, both the perceived value and sense of reality of live commerce were found to positively influence purchase intentions. This is consistent with the results of a study by Oh and Shin, which explored the impact of the behavioral support and inherent characteristics of live commerce platforms on perceived shopping value and purchase intentions [21]. Third, customer satisfaction was found to have a positive impact on purchase intentions. Finally, customer satisfaction was identified as a mediator between perceived value, sense of reality, and purchase intentions.

In the recent consumer market, there have been significant changes in business models focusing on digital transformation. Thus, this study can be considered timely in examining the domestic live commerce landscape. The practical implications for companies operating live commerce platforms include the following. First, the utilitarian value of live commerce was found to have the greatest impact on purchase intentions. Additionally, hedonic value also positively influences purchase intentions. Hence, companies should strive to ensure that consumers perceive not only utilitarian value but also hedonic value when using live commerce. Second, as e-commerce becomes widespread, trust is crucial for service providers and users. To enhance purchase intentions in live commerce, companies need to engage in continuous communication with users, listen to their opinions, and go beyond one-time product sales in their marketing efforts.

From an academic perspective, this study contributes by proposing a new model that adds customer satisfaction as a mediating parameter to the existing relationship path between perceived value and purchase intention. By identifying a new path leading to purchase intention, the study distinguishes itself from previous research.

Despite these practical and academic implications, the study has certain limitations. There is a lack of extensive prior research in the domestic live commerce market, and efforts to evenly distribute age groups for generalization resulted in 40 ~ 50-year-olds comprising 60% of the sample, preventing a thorough analysis of the purchase intentions of the primary target audience of live commerce. Future research should consider incorporating the opinions of the key target demographic, the 20–30 age group, to analyze factors important to them and explore meaningful differences between groups.

Given the recent growth of live commerce facilitated by advancements in interaction and individual recommendation system algorithms, the study proposes future research tasks focusing on consumer attitudes and empirical studies on consumer trust in new technologies such as metaverse, built on big data and AI. Additionally, if subsequent research delves into virtual consumption cultures based on the properties of virtual platforms, it could yield meaningful results regarding future consumer lifestyles and values.

References

1. Arora A, Glaser D, Kim A, Kluge P, Kohli S, Sak N (2021) It's showtime! How live commerce is transforming the shopping experience. McKinsey Insights 21:1–8

2. Mun SR (2023) Live commerce has emerged as a trend of distribution beyond growth. Cosmetic Mania News, 2023/4/27. https://www.cmn.co.kr/sub/news/news_view.asp?news_idx=41467
3. Wongkitrungrueng A, Assarut N (2020) The role of live streaming in building consumer trust and engagement with social commerce sellers. J Bus Res 117(Sep):543–556
4. Dodds William B, Monroe KB, Dhruv G (1991) The effects of price, brand and store ınformation on buyers' product evaluations. J Mark Res (JMR) 28(3):307–319
5. Kim SD, Huang JY, Yang SB (2021) Factors influencing live commerce consumers' watching and purchase intentions: focusing on the Taobao platform in China. J Internet Electron Commerce Res 21(2):53–78
6. Rintamäki T, Kanto A, Kuusela H, Spence MT (2006) Decomposing the value of department store shopping into utilitarian, hedonic and social dimensions: evidence from Finland. Int J Retail Distrib Manage
7. Babin BJ, Darden WR, Griffin M (1994) Work and/or fun: measuring hedonic and utilitarian shopping value. J Consumer Res 20(3):644–656
8. Biocca F, Levy MR (eds) (1995) Communication in the age of virtual reality, 1st edn. Routledge, New York
9. Gao W, Li X (2019) Building presence in an online shopping website: the role of website quality. Behav Inf Technol 38(1):28–41
10. Ou CX, Pavlou PA, Davison RM (2014) Swift guanxi in online marketplaces: the role of computer-mediated communication technologies. MIS Quart 38
11. Li H, Daugherty T, Biocca F (2003) The role of virtual experience in consumer learning. J Consum Psychol 13(4):395–404
12. Huff AS (2000) Changes in organizational knowledge production. Acad Manag Rev 25(2):288–293
13. Oliver RL (1993) Cognitive, affective, and attribute bases of the satisfaction response. J Consumer Res 20(3):418–430
14. Jang SJ, Ryu SK (2022) Platform attribute on satisfaction and repeat-using intention in live commerce: focusing on MZ generation. Korean J Commun Stud 30(3):233–267
15. Dick AS, Basu K (1994) Customer loyalty: toward an integrated conceptual framework. J Acad Mark Sci 22(2):99–113
16. Tsang MM, Ho SC, Liang TP (2004) Consumer attitudes toward mobile advertising: an empirical study. Int J Electron Commer 8(3):65–78
17. Jang SY, Park JK (2020) A study on the moderating effect of narcissism in relationship between perceived consumption value and purchasing intention. J Marketing Manage Res 25(3):63–85
18. Cronin JJ, Brady MK, Hult GTM (2020) Assessing the effects of quality, value and customer satisfaction
19. Choi BN, Han KI (2018) The effects of mobile social commerce characteristics on the purchase intention: focusing on the mediating effects of shopping values. J Digit Convergence 19(8):89–98
20. Oh MO, Shin JG (2021) The effect of behavioral support and source characteristics of live commerce platform on perceived shopping value and purchase intention. J Korean Data Anal Soc 23(5):2247–2265
21. Choe EJ, Jeon SM (2022) What is the influence of sellers on the value of consumers using live commerce and E-WOM intentions? Korea Venture Entrepreneurship Association: Academic Conference Papers, pp 135–139

Analysis of the Impact of Active Parental Intervention, Digital Literacy, Online Opportunities, and Self-regulation on Adolescents' Online Usage

Eui-Yeon Na and Jung-Ah Choi

Abstract Since 2020, the world has been experiencing significant disruptions due to the COVID-19 pandemic. The pandemic has brought about many changes in daily life, leading to the rapid advancement of digital technology. There is a growing interest in digital literacy and the ability to leverage digital skills. Rather than worrying about the risks for our children in the accelerated digitalization of daily life, it is crucial to focus on instilling proper digital literacy and self-regulation skills. This will enable our children to benefit from the various opportunities offered by a digitally based lifestyle. The purpose of this study is to derive the impact of active parental intervention, digital literacy, online opportunities, and self-regulation on adolescents' online usage. To achieve this goal, an online survey using Instagram was conducted among middle and high school students from February 26 to March 12, 2022, for a period of two weeks. A total of 469 responses were collected and used for empirical analysis. SPSS and Smart PLS programs were employed for the analysis. The research findings revealed that active parental intervention has a static impact on digital literacy, online opportunities, and self-regulation. Digital literacy positively influences online opportunities and self-regulation, while self-regulation enhances online opportunities. Lastly, the relationship between active parental intervention and online opportunities shows that digital literacy and self-regulation mediate this relationship. This study is significant because it highlighted the importance of examining the influence of digital literacy and adolescents' self-regulation abilities in the relationship between parental intervention and online opportunities.

Keywords Active parental intervention · Digital literacy · Online opportunities Adolescents' self-regulation · Online opportunities · Internet utilization ability

E.-Y. Na (✉)
Korea Smart Future Education Association, Seoul, Republic of Korea
e-mail: naa0710@naver.com

J.-A. Choi
Department of Program in Project Management, Soong Sil University, Seoul, South Korea
e-mail: lemontree007@soongsil.ac.kr

R. Lee et al. (eds.), *AI and Metaverse*, Studies in Computational Intelligence 1160,
https://doi.org/10.1007/978-3-031-61163-6_4

1 Introduction

The transformation of modern society into a digital society is no longer a choice but a necessity. Rather than worrying about the risks for our children in the era of information intelligence, it is essential to focus on instilling proper digital literacy and self-regulation skills. This will enable them to enjoy various benefits and opportunities in the digitally based lifestyle.

While adolescents are generally considered to be at the forefront of digital technology as digital natives, it cannot be assumed that they possess adequate digital skills [13, 18]. According to research, adolescents face difficulties in finding, managing, and evaluating necessary information, as well as in managing their personal information online and ensuring their safety [4].

Until now, there has been minimal prior research on parental mediation guiding parents to develop online opportunities for their children. Most parental mediation studies have focused on managing internet addiction and online risks. As the digital environment becomes commonplace in daily life, there is a need for research on parental mediation that helps avoid online risks and enhances online opportunities.

This study narrows down the focus to active parental intervention based on previous research suggesting that it can reduce online risks. Thus, this research aims to explore not only the impact of parents' active intervention and digital literacy on online opportunities but also the mediating mechanisms between active parental intervention, digital literacy, and online opportunities. This approach seeks to provide new insights into how adolescents can beneficially utilize the digital environment and how parents can play an appropriate role in this context.

This study is anticipated to contribute insights into the mechanisms behind the role of active parental intervention and factors that can increase online opportunities. The findings can offer solutions for enhancing digital literacy strategies among adolescents who are exposed to various online opportunities.

Ultimately, this research aims to analyze the impact of active parental intervention, digital literacy, online opportunities, and self-regulation comprehensively and empirically on adolescents' online usage.

2 Theoretical Background

2.1 Previous Studies on Parental Mediation

While many previous studies on parental mediation of media content have focused on more traditional media such as television [14], there is a growing emphasis on research related to the internet and parental strategies [8]. Unlike the significant attention given to video games, there is scarce research specifically focusing on tablets and smartphones [8].

Many previous studies propose parental internet mediation to reduce the risks of Problematic Internet Use (PIU). This is due to the concern of parents regarding excessive and uncontrolled internet use, obsession with the internet, and the negative impact on academic performance and interpersonal relationships [15]. Positive results in reducing the risk of PIU through parental mediation have been found in numerous studies [3, 12, 15]. Recent research has shown significant interest in whether parental mediation of social media usage is associated with adolescents' actual social media use.

2.2 Previous Studies on Digital Literacy

Due to the importance of digital literacy (DL), numerous studies related to DL can be found in the literature. Most research has taken a general perspective on students' DL, traditionally investigating the correlation between DL and the mentioned constructs across the entire sample. However, these studies often overlook the potential variations in the availability and use of Information and Communication Technology (ICT) among student subgroups based on their background differences [6].

Most relevant studies have focused on how overall assessment scores related to background variables, primarily relying on either Classical Test Theory (CTT) or Item Response Theory (IRT). These analyses tend to lack fine distinctions between various dimensions of DL [14, 23].

In contrast, a minority of studies have adopted a multidimensional IRT model. For example, Siddiq et al. [22] used a multidimensional Rasch model to analyze a learning-ICT literacy test in the context of digital networks across four dimensions [22]. This approach provides detailed diagnostic information, revealing students' strengths and weaknesses in each technical dimension of DL. It can be utilized to inform teaching strategies, enhance student learning, and explore the relationship between students' background variables and the mastery or non-mastery of each digital skill.

2.3 Previous Studies on Self-Regulation

Self-regulation refers to the ability to control oneself, maintaining regularity and adherence to certain standards. This ability involves regulating one's thoughts, emotions, and behaviors. The development of self-regulation progresses autonomously and progressively from early childhood, and various models have been developed to explain the regulatory processes within the social environment as individuals age.

Examining various studies on self-regulation, the concepts of 'self-regulation' and 'self-regulation' are often considered similar. Self-regulation is generally conceptualized as a multidimensional construct that includes emotions, cognition, and behavior

[19]. The core of self-regulation is the belief that individuals who are aware of their goals can control and direct their own behavior. While cognitive mechanisms form a crucial part of self-regulation, achieving desired outcomes requires cognitive skills beyond mere cognition. Self-regulation is broadly defined as a repertoire of cognitive, metacognitive, emotional, and motivational factors that allow individuals to monitor and adjust their thoughts, emotions, and behaviors [17]. The ability to evaluate the effectiveness of applied cognitive strategies, observe, and use emotions experienced during activities, and modify the adopted approach when necessary is essential when exploring initiated activities. However, these non-cognitive components of the creative process are often overlooked [11]. In this study, the influence of these non-cognitive components on self-regulation is distinctive.

2.4 Previous Studies on Online Opportunities

The development of information technology has led to the widespread use of various information devices, and children worldwide primarily access the internet for information, entertainment, connecting with friends, and participating in communities, mainly using mobile devices. The increasing use of the internet and devices in society serves as a factor in reducing the gap in internet access experienced by various groups of children. In situations with low accessibility, children take the lead in adopting new technologies and connect more online than adults.

Previous research has suggested various opportunities in online communication, such as enhancing self-esteem, increasing opportunities for peer relationship formation, improving the quality of existing relationships, and enhancing opportunities for sexual self-exploration. However, certain risk factors, including cyberbullying and sexual online solicitation, have also been clearly identified. According to research, the internet is not a place where adolescents solely face insurmountable risks, nor does it automatically create opportunities for positive psychosocial development [24].

The effects of online communication on psychosocial development can vary depending on technological, situational, and individual factors. However, our review suggests that a proper understanding of adolescents' psychosocial development cannot ignore online communication any longer. The argument that media enhances the potential for adolescents' self-socialization is not new [1]. However, adolescents have never had the opportunity to explore their identity through such diverse means without the supervision of traditional socialization agents such as parents or schools. Overall, our review indicates that while this does not necessarily have to be a cause for concern, it certainly raises important questions about the future roles of parents and schools in the psychosocial development of adolescents Fig. 1.

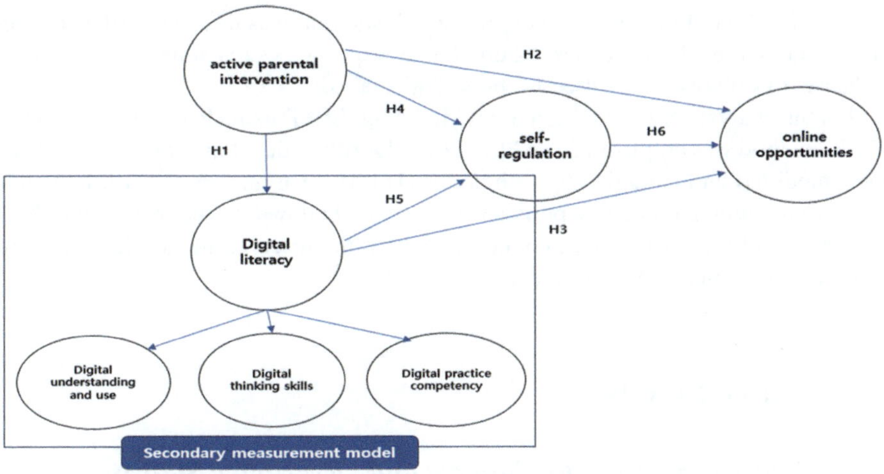

Fig. 1 Research model

3 Research Method

3.1 Research Model and Hypotheses

3.2 Hypothesis

In this study, we propose the following hypothesis to determine the relationship between active parental intervention, digital literacy, and adolescents' online opportunities, and whether self-regulation has a mediating effect on each variable.

H1. Active parental intervention will have a positive impact on digital literacy.

H2. Active parental intervention will have a positive impact on online opportunities.

H3. Digital literacy will have a positive impact on online opportunities.

H4. Active parental intervention will have a positive impact on self-regulation.

H5. Digital literacy will have a positive impact on self-regulation.

H6. Self-regulation will have a positive impact on online opportunities.

3.3 Survey Methods

The survey data were collected from middle and high school students residing in Seoul and Gyeonggi Province from February 26 to March 12, 2022, over a period of two weeks. An online questionnaire through Instagram was utilized for data collection. Among the 531 responses, 62 unreliable responses were excluded, resulting in a dataset of 469 responses used for empirical analysis.

For the analysis, Structural Equation Modeling (SEM) using Partial Least Squares (PLS) method was employed. The PLS-SEM algorithm, developed by Wold [25] and later modified and extended by Lohmöller [16], is a variance-based SEM suitable for research with a predictive purpose. This method allows for the inclusion of both reflective and formative measurement models in the research model. In this study, only reflective measurement models were used.

4 Research Results

4.1 *Characteristics of the Sample and Descriptive Statistics*

This study includes demographic variables such as gender, age, parental age, and education. The distribution of the most frequently used internet activities showed that 195 participants (41.6%) engaged in communication, 167 participants (35.6%) used social media, and 73 participants (15.6%) were involved in movie-watching. Other activities included SNS, gaming, web novels, YouTube, and more. The distribution of daily digital usage time indicated that 7 participants (1.5%) used digital devices for less than 1 h, 39 participants (8.3%) for 1–2 h, 161 participants (34.3%) for 3–4 h, and 262 participants (55.9%) for more than 4 h.

Descriptive statistics for individual measurement variables were calculated using the SPSS 23.0 program. No outliers were identified as the standard deviation was within ± 3 for all measurement variables. Additionally, examining kurtosis and skewness to assess data normality revealed that most values fell within the acceptable range of − 1 to + 1. Although some kurtosis values exceeded + 1, indicating non-normal distribution for those specific items, they were not significantly deviating from normality. Overall, normality was considered satisfactory.

4.2 *Evaluation of the Measurement Model*

To validate the fit of the research model, this study conducted reliability and validity analyses using SPSS 23.0 and SmartPLS 3.0. The analysis proceeded in two steps: (1) exploratory factor analysis using SPSS to verify validity, assess reliability, and analyze correlations, and (2) structural equation modeling (PLS-SEM) with confirmatory factor analysis to assess model fit and validity (discriminant validity, convergent validity).

The analysis results indicated no issues with reliability and validity. Through internal consistency verification for each variable, a total of 29 survey items across 4 variables were retained, excluding six survey items that demonstrated low reliability in the internal consistency test.

(1) Second-order measurement model of digital literacy

In the second-order confirmatory factor analysis of digital literacy, reliability measures such as Cronbach's Alpha, rho_A, and CR all exhibited values exceeding 0.7, external loadings were all above 0.7, and the Average Variance Extracted (AVE) values surpassed the threshold of 0.5. This confirms the reliability of measurement variables and the convergent validity of latent variables.

(2) Internal consistency reliability

The evaluation of internal consistency reliability for latent variables and measurement variables showed that all reliability measures (Cronbach's Alpha, rho_A, CR) exceeded 0.7, indicating desirable reliability. Internal consistency reliability was considered to be established [20].

(3) Convergent validity assessment

The analysis of external loadings for latent variables and measurement variables revealed that all external loadings were above 0.7, confirming convergent validity [2, 10]. Thus, the measurement variables were deemed suitable for use in the measurement model. Both Indicator Reliability and AVE values for individual measurement variables were above the threshold of 0.5, confirming the reliability of individual measurement variables and the convergent validity of latent variables [5, 9].

(4) Discriminant validity assessment

Discriminant validity was assessed using three criteria: HTMT criterion, Fornell-Larcker criterion, and cross-loading analysis. All three criteria confirmed the discriminant validity between latent variables.

4.3 Evaluation of the Structural Model and Hypothesis Testing

After assessing the measurement model, confirming the reliability and validity of the observed and latent variables, the evaluation of the structural model proceeded through five stages.

The first stage involves evaluating the multicollinearity among latent variables. Examining the internal Variance Inflation Factor (VIF) for each endogenous latent variable revealed values consistently below 5, indicating no issues of multicollinearity among the latent variables.

The second stage assesses the explanatory power of exogenous latent variables for endogenous latent variables. The R^2 values signify weak explanatory power when R^2 is 0.25, moderate explanatory power when R^2 is 0.50, and strong explanatory power when R^2 is 0.75. In the results of the Partial Least Squares (PLS) algorithm, the adjusted R^2 for the endogenous latent variable, representing online opportunities, is 0.169, indicating a weak explanatory power.

The third stage focuses on determining the relative impact of exogenous latent variables on endogenous latent variables, represented by the f^2. Small, medium, and large effect sizes for f^2 are evaluated at 0.02, 0.15, and 0.35, respectively. The analysis indicates that the parent intervention contributes with a medium effect size ($f^2 = 0.114$) to the path coefficient of digital literacy, making it the most substantial contribution.

Parental intervention has a small effect size ($f^2 = 0.066$) on self-regulation, while digital literacy contributes with small effect sizes ($f^2 = 0.086$ for online opportunities, $f^2 = 0.058$ for self-regulation).

The fourth stage evaluates whether the structural model possesses predictive relevance for specific latent variables. Predictive relevance (Q^2) measures how well the model predicts the endogenous latent variables. In this study, the Q^2 values for all endogenous latent variables range from 0.053 to 0.102, indicating positive values greater than 0, suggesting that the model has predictive relevance.

The hypothesis testing results indicate that all six hypotheses (H1, H2, H3, H4, H5, H6) have been accepted. In addition, the indirect effects of all paths were found to be established in the relationship between the main factors that verified the indirect effects of digital literacy and self-regulation in the relationship between parents' active intervention and online opportunities Table 1.

5 Conclusion

The purpose of this study is to comprehensively and empirically analyze the impact of parental intervention, digital literacy, online opportunities, and self-regulation on adolescents' online use. Based on the research results, the conclusions drawn are as follows:

First, the structural relationship in which parents' active intervention affects adolescents' online opportunities was found to be a variable that significantly affects not only parental intervention and digital literacy but also self-regulation in the process of learning that was not covered in previous studies. This is a good indication of what was overlooked by the macro-ecological system perspective that providing parental intervention and digital literacy will lead to online opportunities. In a digital environment that enables a wide range of opportunities, including information and education, community collaboration and social relationships, and creative digital content, the ultimate purpose of teenagers' participation in the digital environment is to improve their online opportunities for success in an individual's digital environment, not just for participation. In addition to the digital environment, it is also necessary to design, develop, and operate a digital ecological system that can link online opportunities to beneficial outcomes by considering programs that support customized self-regulation technologies for teenagers who prefer digital environments as well as macro-system programs for parental intervention and digital literacy

Table 1 Evaluation results for significance and fit of hypothesized paths

Path		Original sample (O)	Sample mean (M)	Standard deviation (STDEV)	T statistics (|O/ STDEV|)	*P* value
H1	Active parental intervention - > Digital literacy	0.32	0.323	0.042	7.618	0
H2	Active parental intervention - > Online opportunities	0.124	0.127	0.047	2.622	0.009
H3	Digital literacy - > Online opportunities	0.289	0.29	0.047	6.149	0
H4	Active parental intervention - > Self-regulation	0.249	0.257	0.047	5.349	0
H5	Digital literacy - > Self-regulation	0.234	0.23	0.049	4.815	0
H6	Self-regulation - > Online opportunities	0.134	0.133	0.048	2.801	0.005
Mediating variable relationships	Active parental intervention - > Self-regulation - > Online opportunities	0.033	0.034	0.014	2.354	0.019
	Active Parental intervention - > Digital literacy - > Online opportunities	0.091	0.093	0.019	4.820	0.000
	Active parental intervention - > Digital literacy - > Self-regulation	0.074	0.075	0.019	2.228	0.026
	Digital literacy - > Self-regulation - > Online opportunities	0.031	0.032	0.013	2.349	0.019
	Active parental intervention - > Digital literacy - > Self-regulation - > Online opportunities	0.010	0.010	0.001	2.228	0.026

and self-regulation models for digital manipulation and management that occur in the process of utilizing digital environments.

Also, although it is clear that active parental intervention is a factor affecting adolescents' online opportunities, it may not be related to actually learning how to acquire digital skills from parents.

According to the results of this study, online opportunities have a direct impact on relationships with digital literacy, self-regulation, and active parental intervention. Specifically, digital literacy has the greatest impact on online opportunities, followed by self-regulation and active parental intervention. These findings somewhat align with previous research that suggests active parental intervention may not necessarily be related to acquiring digital skills from parents [7, 21].

Active parental intervention involves encouraging adolescents to use digital media constructively by answering their questions, explaining media content, and discussing its appropriateness [21]. For example, proactive parental intervention practices, such as teaching children how to safely use social media in the context of digital technology, can enhance adolescents' critical thinking and self-judgment abilities, thereby improving self-regulation. However, as proactive parental intervention does not necessarily involve actively teaching digital skills, there is a need for new parental intervention methods tailored to digital literacy and online opportunities.

Secondly, adolescents' online opportunities are not only directly influenced by active parental intervention but also indirectly affected through the mediation of digital literacy and self-regulation. Even if parental intervention skills are high, if adolescents do not cultivate the ability to extract beneficial results from the digital environment for themselves and others and do not develop the skills to select, execute, and evaluate digital resources, their ability to increase their online opportunities may be limited. This emphasizes the importance of the pedagogical aspects of digital learning, highlighting the need for attention to the process of digital learning that internalizes knowledge, skills, and attitudes related to self-regulation predicted by digital literacy.

Thus, parents, educational institutions, and government agencies need new ways to enhance parental intervention skills and foster adolescents' online opportunities within the diverse environments of the ecological system. This requires considering the beneficial interactions at various levels (micro, meso, exo, macro, and time system levels) within the ecological system to ensure effective operation. It is essential to develop new methods that can increase parental intervention skills and promote new ways for adolescents to grow their online opportunities. This underscores the need for developing and operating online opportunity programs based on adolescent digital literacy, considering factors that can positively express the self-regulation characteristics of adolescents that are easily manifested in the digital learning process.

Finally, the impact relationship on adolescents' online opportunities investigated in this study is challenging to generalize cause and effect due to the examination of a single point in time. Additionally, since the influence of parental intervention, digital literacy, and self-regulation on online opportunities is more likely to be formed over a relatively medium to long-term period rather than in the short term, future research should focus on longitudinal studies rather than cross-sectional studies.

References

1. Arnett JJ (1995) Adolescents' uses of media for self-socialization. J Youth Adolesc 24(5):519–533
2. Bagozzi RP, Yi Y, Phillips LW (1991) Assessing construct validity in organizational research. Admin Sci Quart 421–458
3. Bleakley A, Ellithorpe M, Romer D (2016) The role of parents in problematic internet use among US adolescents. Media Commun 4(3):24–34
4. Calvani A, Fini A, Ranieri M, Picci P (2012) Are young generations in secondary school digitally competent? A study on Italian teenagers. Comput Educ 58(2):797–807
5. Chinn WW (1998) The partial leat squares approach to structural equation modeling. Mod Methods Bus Res 295(2):295–336
6. Chinn MD, Fairlie RW (2010) ICT use in the developing world: an analysis of differences in computer and internet penetration. Rev Int Econ 18(1):153–167
7. Clark LS (2011) Parental mediation theory for the digital age. Commun Theory 21(4):323–343
8. Connell SL, Lauricella AR, Wartella E (2015) Parental co-use of media technology with their young children in the USA. J Child Media 9(1):5–21
9. Fornell C, Larcker DF (1981) Structural equation models with unobservable variables and measurement error: algebra and statistics
10. Hair JF, Ringle CM, Sarstedt M (2011) PLS-SEM: indeed a silver bullet. J Market Theory Practice 19(2):139–152
11. Ivcevic Z, Brackett MA (2015) Predicting creativity: Interactive effects of openness to experience and emotion regulation ability. Psychol Aesthet Creat Arts 9(4):480–487
12. Kalmus V, Blinka L, Olafsson K (2015) Does it matter what mama says: evaluating the role of parental mediation in European adolescents' excessive Internet use. Child Soc 29(2):122–133
13. Li Y, Ranieri M (2010) Are "digital natives" really digitally competent?—A study on Chinese teenagers. Br J Edu Technol 41(6):1029–1042
14. Livingstone S, Helsper EJ (2008) Parental mediation of children's internet use. J Broadcast Electron 52(4):581–599
15. Livingstone S, Olafsson K, Helsper EJ, Lupianez-Villanueva F, Veltri GA, Folkvord F (2017) Maximizing opportunities and minimizing risks for children online: the role of digital skills in emerging strategies of parental mediation. J Commun 67:82–105
16. Lohmöller JB (1989) Predictive versus structural modeling: Pls versus ml. In: Latent variable path modeling with partial least squares. Physica, Heidelberg, pp 199–226
17. Malanchini M, Engelhardt LE, Grotzinger AD, Harden KP, Tucker-Drob EM (2019) Same but different: associations between multiple aspects of self-regulation, cognition, and academic abilities. J Pers Soc Psychol 117(6):1164–1188
18. Margaryan A, Littlejohn A, Vojt G (2011) Are digital natives a myth or reality? University students' use of digital technologies. Comput Educ 56(2):429–440
19. McClelland MM, Ponitz CC, Messersmith EE, Tominey S (2010) Self-regulation: integration of cognition and emotion
20. Nunnally B, Bernstein IR (1994) Psychometric theory. Oxford University, New York
21. Rodríguez-de-Dios I, van Oosten JM, Igartua JJ (2018) A study of the relationship between parental mediation and adolescents' digital skills, online risks and online opportunities. Comput Hum Behav 82:186–198
22. Siddiq F, Gochyyev P, Wilson M (2017) Learning in digital networks–ICT literacy: a novel assessment of students' 21st century skills. Comput Educ 109:11–37
23. Siddiq F, Scherer R (2019) Is there a gender gap? A meta-analysis of the gender differences in students' ICT literacy. Educ Res Rev 27:205–217
24. Valkenburg PM, Peter J (2011) Online communication among adolescents: an integrated model of its attraction, opportunities, and risks. J Adolesc Health 48(2):121–127
25. Wold H (1982) Models for knowledge. The making of statisticians, pp 189–212

A Study on Housing Preference Characteristics of Single-Person Households—Focusing on a Comparison Between Seoul's Youth and Middle-Aged People-

Yoonhee Kim, Gwangyong Gim, Jungsun Oh, and Jeeyoeng Kim

Abstract In South Korea, the number of single-person households has been rapidly increasing due to factors such as the aging population, low birth rates, changing perceptions of marriage, and family dissolution. The growth of single-person households is an inevitable social structural change, observed not only in advanced countries but also in relatively fast-growing economies. This study aimed to investigate the characteristics of single-person household members residing in Seoul, focusing on middle-aged individuals (40–64 years old) and young adults (under 40 years old). Additionally, the research analyzed how housing selection factors and preferred housing types differ across age groups. Specifically, an online survey was conducted to examine the key characteristics of single-person households among middle-aged and young adults living in Seoul, and a t-test was employed to compare housing preferences between the two groups. The results indicated that the younger age group prioritizes factors such as crime prevention, accessibility to public transportation, good housing conditions, proximity to the workplace, nearby convenience facilities, and safety. In contrast, individuals aged 40 and above in the middle-aged group prioritize factors like the presence of nearby parks, educational environments, and parking facilities, with a strong emphasis on green living environments. This research underscores the need for tailored housing policies for single-person households based on household types and age-related housing demand considerations.

Keywords Single · Person households · single · Person households satisfaction with housing · single · Person households for middle · Aged people

---

Y. Kim (✉) · J. Oh · J. Kim
Department of IT Policy and Management, Soongsil University, Seoul, South Korea
e-mail: kunice70@naver.com

G. Gim
Department of Business Administration, Soongsil University, Seoul, South Korea
e-mail: gygim@ssu.ac.kr

R. Lee et al. (eds.), *AI and Metaverse*, Studies in Computational Intelligence 1160,
https://doi.org/10.1007/978-3-031-61163-6_5

1 Introduction

In 2019, single-person households in Seoul accounted for 33.3% of all households, making it the most predominant household type. Statistics Korea (2019) forecasts that the upward trend of single-person households in Seoul will reach 37.2% of all households by 2047. Living alone has evolved from being something special to becoming the predominant household type. Single-person households have become a core component of the future population structure in South Korea, especially among middle-aged individuals. The increase in middle-aged single-person households is expected to continue due to increased life expectancy, changing family structures, and ongoing growth. It is anticipated that this rise in middle-aged single-person households is not a transitional phenomenon but will become a prevalent household type [1, 2]. Over the next 30 years, household downsizing is expected to persist in South Korea. While the number of single-person households among young adults is predicted to decrease by age group, single-person households among middle-aged individuals are expected to increase as they gradually enter old age. It is estimated that the number of single-person households in their 60s will be twice the current figure, and those in their 70s will be more than three times the current size [3].

This study focuses on middle-aged (40–64 years old) and young adult (under 40 years old) single-person households residing in Seoul. It aims to investigate the characteristics of single-person household members, analyze the housing selection factors and preferred housing types, and identify the differences between middle-aged and young adult single-person households. Satisfaction with housing can vary depending on factors such as age, income, gender, housing type, location, environment, occupancy, and community. Therefore, detailed and integrated housing policies are needed to enhance housing satisfaction for single-person households.

This study seeks to address the concern that the growing population of middle-aged single-person households is expected to transition naturally into the elderly population. As a result, issues related to single-person households in the elderly demographic may become more severe in the future. Single-person households exhibit varying characteristics, and the factors influencing housing satisfaction differ in magnitude. It is expected that the housing preferences of middle-aged single-person households will differ from those of young adult single-person households, including housing types and features. Investigating and analyzing the housing types preferred by middle-aged single-person households will help them prepare for their future retirement and later life. The results of this study emphasize the need for tailored housing policies to meet diverse housing demands by generation and household type. Thus, the research questions derived from this study are as follows.

Research Question 1. How does housing satisfaction among single-person households rank? What are the top 3 and bottom 3 factors contributing to housing satisfaction?

Research Question 2. What are the differences in housing preference types between middle-aged and young adult single-person households?

2 Theoretical Background

2.1 Domestic Single-Person Household Status and Housing Types

Single-person households are mixed with single or single households, and the National Statistical Office classifies households that live alone, that is, households that make a living independently, such as cooking and sleeping, as single-person households. It is characterized by the increasing tendency of late marriage non-marriage, the increasing rate of divorce, and the increasing proportion of men in middle-aged single-person households and elderly single-person households. According to the National Statistical Office's announcement, the types of single-person households in Seoul in 2019 were detached houses (40.4%), apartments (22.1%), and multi-family houses (17.2%), and the ratio of non-housing (officetels, Gosiwon, lodging rooms, etc.) was 15.4%, showing various types of housing. In terms of housing type, single-person households have the highest percentage of single-person households living in single-family houses in all age groups, and multi-person households have the highest percentage of living in apartments until young, middle-aged, and elderly people. In all age groups of single-person households, the proportion of detached houses is high, but the younger the population, the higher the proportion of officetel dwellings, and the higher the proportion of apartment dwellings [4]. Single-person households predominantly prefer monthly rent over security deposit (jeonse). Among these households, there is a clear inclination towards residing in officetels for young adults, while middle-aged single-person households exhibit a strong preference for apartments [5]. Regarding housing type preferences among single-person households, it was observed that, as income increases, there is a tendency for them to reside in high-density, shared housing [6]. When examining the characteristics of single-person households, they are motivated by factors like cost-effectiveness, avoiding social isolation, and prioritizing accessibility to urban facilities, such as shops and restaurants. In the case of young adults, despite the challenges they face in terms of financial stability, they divide into two distinct groups. One group seeks economic stability and opts to form single-person households, while the other group, despite economic instability, chooses an independent lifestyle. Young adult single-person households exhibited a significantly higher proportion of rental housing (over 50%). However, their residential environments did not substantially differ from households with more members [7].

Single-person households tend to concentrate in areas with convenient access to city centers. There are significant gender imbalances observed between neighborhoods. For example, in areas like Gwanak-gu, where officetel communities are well-developed, there is a notable male dominance. In contrast, Gangnam-gu, a commercial and business hub, shows a distinct female dominance [8]. While typical households tend to have low residential mobility, single-person households, especially among young adults, exhibit significantly higher rates of residential mobility, with 62% of young adult single-person households and 40% of middle-aged single-person households moving residences. This higher mobility indicates relatively lower housing stability for single-person households [9]. The residential mobility of single-person households displays distinct characteristics when compared to other household types in terms of age groups and geographic locations. These unique features include lower reliance on car usage compared to regular households, a preference for residing in areas with well-developed public transportation access, such as transit-oriented developments, and a strong inclination towards living in close proximity to workplaces [10].

2.2 *Preferred Housing Types for Single-Person Households*

Residential satisfaction is a subjective assessment of the dwelling one occupies, influenced by familial, personal, and cultural factors, impacting housing choice and subsequent satisfaction with the living space [11]. Examining the housing preferences affecting the satisfaction of the elderly, it is noted that males from the baby boomer generation tend to express a higher inclination towards townhouses and rural homes, while females tend to favor apartments for their convenience in household management and chores [12]. In studies on the life satisfaction based on the types of residences among the elderly, it was observed that although a significant number of seniors reside in single-family homes as opposed to apartments, those living in apartments reported higher life satisfaction [13]. The housing space for the elderly, whether in apartments or standalone houses, significantly impacts their overall satisfaction with life. This extends beyond the interior features of the housing to include the characteristics of the community and the convenience of nearby facilities, all of which are directly related to the satisfaction and contentment of the elderly [14]. Furthermore, factors such as proximity and satisfaction with commercial facilities, cultural amenities, educational environments, city parks, green spaces, parking facilities, security conditions, and air quality influence overall residential satisfaction [15]. Specifically, the housing types preferred by voluntary single-person households in the middle to later stages of life include apartments or upscale villas. They tend to prefer units with over 50 households, lower than 10 floors, and require additional amenities such as exercise spaces and cafes. There is a preference for housing units with two or more rooms, especially a desire for an expanded living room area. The tendency among this group is to own properties of 20 square meters or more, while perceiving maintenance costs in the range of 100,000 to 300,000 Korean won per month as reasonable.

These results indicate a distinctive difference in environmental characteristics compared to existing residential options for single-person households, such as multi-family residences, urban lifestyle homes, and studio apartments. This suggests the necessity for the development of new housing models tailored for voluntary single-person households in the middle to later stages of life [16].

3 Research Method

This study employed a theoretical research approach based on literature review and analysis through a survey. Residential satisfaction was measured using a Likert 5-point scale. The population and sample selection involved 432 participants, including single-person households aged 20 and above residing in urban and southeastern districts of Seoul. The survey was conducted in September 2021. The survey focused primarily on the growing demographic of individuals in their 40s, with a significant portion of the sample allocated to this age group. Participants were randomly surveyed through online and mobile platforms by Embrain, a survey agency. Stratified sampling was employed based on residential areas, age groups, household types, and age distribution to ensure representation (Table 1).

Table 1 Demographic statistics

Variable	Sortation	Frequency	Percentage
Gender	(1) Male	158	36.6
	(2) Female	274	63.4
Age	(1) 20–29 years old	42	9.7
	(2) 30–39 years old	80	18.5
	(3) 40–49 years old	199	46.1
	(4) 50–59 years old	80	18.5
	(5) Over 60 years old	31	7.2
Housing type	(1) Apartment	207	47.9
	(2) Single-family house	47	10.9
	(3) Villa/Joint	136	31.5
	(4) Officetel	29	6.7
	(5) Other	13	3
Household type	(1) General	280	64.8
	(2) Single-person household	152	35.2
Reasons for single-person households	(1) Bereavement	3	0.7
	(2) Divorce	21	4.9
	(3) Separate	3	0.7
	(4) Single	116	26.8
	(5) Other	9	2.1

4 Research Result

4.1 Priority in Selecting Residential Areas by Age Group

To investigate differences in priority when selecting residential areas, an independent sample t-test was conducted. The results indicated that the younger age group (M = 4.39, SD = 0.637) showed significantly higher preference for accessibility to public transportation compared to the middle-aged group (M = 4.22, SD = 0.630) ($p < 0.01$). Additionally, the preference for living in proximity to workplaces was significantly higher in the younger age group compared to the middle-aged group ($p < 0.05$). Similarly, the prioritization of nearby amenities showed a significant preference among the younger age group compared to the middle-aged group (p < 0.05). On the other hand, the middle-aged group showed significantly higher priority for factors like parking facilities and neighborhood environment compared to the younger age group ($p < 0.01$). Moreover, the preference for educational environments was significantly higher in the middle-aged group compared to the younger age group ($p < 0.05$) (Table 2).

Table 2 Priority in selecting residential areas (by age group)

	Group	Average	Standard deviation	t	*p*
Accessibility to public transportation	Young	4.39	0.637	2.626	0.009**
	Middle-aged	4.22	0.630		
Good condition of residence	Young	4.35	0.642	1.879	0.061
	Middle-aged	4.22	0.671		
A place close to one's job	Young	4.19	0.660	2.070	0.039*
	Middle-aged	4.03	0.743		
The surrounding facilities	Young	4.25	0.659	2.286	0.023*
	Middle-aged	4.09	0.625		
Educational environment	Young	3.35	1.020	− 2.120	0.035*
	Middle-aged	3.57	0.962		
Security and safety environment	Young	4.43	0.691	1.396	0.163
	Middle-aged	4.32	0.696		
Parking	Young	3.65	1.036	− 3.044	0.003**
	Middle-aged	3.97	0.839		
The surrounding environment such as the neighbors	Young	3.55	0.844	− 2.863	0.005**
	Middle-aged	3.80	0.754		

* $P < 0.05$, ** $p < 0.01$, *** $P < 0.001$

Table 3 A survey of housing satisfaction of single-person households

Priority when choosing a residence	Ranking	Average
Traffic conditions	1	3.99
Direct proximity	2	3.83
Medical health facilities	3	3.82
Access to urban parks and green areas	4	3.64
Security and crime prevention	5	3.58
Cultural and welfare facilities	6	3.57
Living amenities and facilities	7	3.51
Educational environment	8	3.50
Future value of investment	9	3.30
Sports facilities	10	3.22
Condition of the house	11	3.22
Housing costs	12	3.20
Housing and living conditions	13	3.17
Relationship with neighbors	14	2.99

4.2 Housing Satisfaction Survey of Single-Person Households

The following are the rankings of single-person households in the survey on housing satisfaction (Table 3).

4.3 Independent Sample t-test on Residential Satisfaction Ranking Between Young and Middle-Aged Single-Person

An independent sample t-test was conducted to examine whether there are differences in the factors considered when choosing residential areas between young and middle-aged single-person households (Table 4).

5 Conclusion and Implications

5.1 Conclusion

This study conducted a comparative analysis of the priority in housing choices influencing the most influential factors for a happy life among general households, single-person households, and different age groups within single-person households. This

Table 4 Priority in residential area selection for young and middle-aged single-person households

		N	Average	Standard deviation	Standard error of the mean
Housing and living conditions	Young	36	2.97	0.941	0.157
	Middle-aged	116	3.23	1.041	0.097
Direct proximity	Young	36	3.94	0.924	0.154
	Middle-aged	116	3.79	0.991	0.092
Condition of the house	Young	36	3.36	0.990	0.165
	Middle-aged	116	3.18	1.027	0.095
Medical health facilities	Young	36	3.75	0.906	0.151
	Middle-aged	116	3.84	0.780	0.072
Traffic conditions	Young	36	3.94	0.984	0.164
	Middle-aged	116	4.01	0.909	0.084
Educational environment	Young	36	3.36	0.867	0.144
	Middle-aged	116	3.54	0.727	0.068
Future value of investment	Young	36	3.22	1.098	0.183
	Middle-aged	116	3.33	0.949	0.088
Cultural and welfare facilities	Young	36	3.64	0.931	0.155
	Middle-aged	116	3.54	0.898	0.083
Access to urban parks and green areas	Young	36	3.17	1.108	0.185
	Middle-aged	116	3.79	0.829	0.077
Relationship with neighbors	Young	36	2.75	1.079	0.180
	Middle-aged	116	3.07	0.743	0.069
Security and crime prevention	Young	36	3.36	1.046	0.174
	Middle-aged	116	3.65	0.897	0.083
Housing costs	Young	36	2.81	1.064	0.177
	Middle-aged	116	3.33	0.985	0.091
Sports facilities	Young	36	3.08	1.052	0.175
	Middle-aged	116	3.27	0.838	0.078
Living amenities and facilities	Young	36	3.42	1.156	0.193
	Middle-aged	116	3.54	0.936	0.087

trend is projected to continue rapidly in the future. Single-person households, driven by reasons like remaining single, being divorced, separated, or affected by the aging population, are notably on the rise, especially among the middle-aged demographic.

Survey results on the primary considerations when choosing residential areas showed that individuals under 40 prioritize access to public transportation, overall condition of the residence, proximity to workplace, nearby amenities, and security more than the middle-aged group. Meanwhile, for those aged 40 and above, factors such as nearby parks or green spaces, educational environment, parking facilities,

Table 5 Areas requiring improvement (in Seoul's housing policies)

	Total participants		Single-person households	
	Frequency	Percentage(%)	Frequency	Percentage(%)
Housing rental policies	122	28.2	48	31.6
Housing financial support policies	136	31.5	54	35.5
Housing improvement support policies	34	7.9	4	2.6
Residential welfare policies	72	16.7	25	16.4
Other	7	1.6	2	1.3
Total	371	85.9	133	87.5

and neighborhood surroundings were given higher priority compared to the younger demographic. Providing urban housing tailored for single-person households aids in preventing urban sprawl and slum formation, creating environments for close proximity to workplaces, thereby contributing positively to urban regeneration. Rejuvenating urban housing supply can prevent population outflow and stagnation in urban functionality, ensuring a resident population in the city. Support is crucial across all domains for both the younger and older generations, including housing, safety, income, caregiving, and social relationships. Addressing housing, one of the most fundamental aspects of life, is particularly urgent (Table 5).

5.2 *Implications*

Regarding the research on "How is the residential satisfaction of single-person households? (in the top 1–3 and bottom 1–3 categories)", the following implications were found. In terms of residential satisfaction, single-person households rank transportation conditions, proximity to workplaces, and healthcare conditions as the top 1–3 priorities. However, housing costs and the quality of housing conditions, as well as the relationship with neighbors, appear in the bottom 1–3 categories, indicating a decreased level of satisfaction with the current living conditions for single-person households. This suggests that single-person households might not experience the same quality of life as general households. Regarding the research on "What are the differences in housing preference types between middle-aged and young age groups?", the following implications were found. The research findings suggest that middle-aged individuals place a higher value on factors like parking facilities and neighborhood surroundings compared to the younger demographic.

As individuals age, their residential satisfaction is influenced not only by personal characteristics but also by psychological and local community environmental factors. Sustained relationships with neighbors and maintaining social connections within the area where one has lived lead to higher residential satisfaction. Consequently, while younger demographics emphasize housing costs, the characteristics of housing demand for the middle-aged prioritize the quality of housing over cost considerations.

Since the survey sample was confined to specific regions within Seoul—Jongno-gu, Jung-gu, Yongsan-gu, Seocho-gu, Gangnam-gu, Songpa-gu, and Gangdong-gu—the statistical insights from the questionnaire may have limitations in representing the entire population. Therefore, future studies are needed to encompass diverse regions and demographics for a comprehensive understanding of the primary characteristics of single-person households and housing types.

To overcome the limitations of the research, improvements in subsequent research should involve an in-depth study on insecure housing environments for middle-aged single-person households, and the inclusion of additional latent factors related to single-person household characteristics for more advanced research outcomes. Continuous research focusing on the housing needs of single-person households and the current status across various regions is expected to enhance housing support policies for this demographic.

References

1. Yuh W (2003) Consumption patterns of the elderly couple and elderly single. J Fam Better Life 21(5):1–12
2. Yeung WJJ, Cheung AKL (2015) Living alone: one-person households in Asia. Demogr Res 32:1099–1112
3. Park M, Lee J (2018) Housing policy responding to one-person households increase. Korea Research Institute for Human Settlements Brief, pp 1–8
4. Shin S (2010) A study on the spatial distribution of one person households the case of Seoul. J Korea Plan Assoc 45(4):81–95
5. Wulff M (2001) Growth and change in one person households: implications for the housing market. Urban Policy Res 19(4):484–486
6. Kim J, Yoo S (2021) Perceived health problems of young single-person households in housing poverty living in Seoul, South Korea: a qualitative study. Int J Environ Res Public Health 18(3):1067
7. Kang E, Lee M (2016) Single-person households in South Korea and their policy implications, vol 2016, issue 4. Korea Institute for Health and Social Affairs, pp 47–56
8. Lee D (2012) Living alone: one-person households in Asia, vol 109. Seoul Studies, pp 1–24
9. Park M, Cho Y (2020) The increasing trend of single-person households by age and gender, and policy responses based on housing characteristics. Korea Research Institute for Human Settlements Brief, pp 1–8
10. Byun M, Min B, Park M (2019) An empirical analysis of the spatial distribution and flow patterns of Seoul's single-person households. Korea J Population Stud 42(4):91–119
11. Kwon S, Park H (2014) Analysis on affecting factors for the housing environment satisfaction of one-person households in Seoul. Hous Stud 22(1):77–104
12. Lim K (2015) A study on residential choice of Korean Baby boomer after retirement. Hansung University, pp 133–134
13. Jeong J (2013) The factors affecting on elderly' happiness in the residential environments. J Regional Assoc Architect Inst Korea 15(6):93–100
14. Koo H, Cho H (2016) An empirical study on housing preferences by awareness of housing and life type of Korean Baby-boomers. J Consumer Stud 27(2):57–84

15. Kim B (2019) The effects of residential characteristics of single person households on residential satisfaction. Korean Urban Manage Assoc 32(1):69–80
16. Seo Y, Kang S (2019) A basic study on the preferred residential space by the characteristics of single-person households-focusing on the middle-ages and voluntary single-person household. In: Autumn annual conference of AIK proceedings, vol 39, issue 2, pp 53–56

A Study on Factors Affecting Seniors' Intention to Use Smart Devices

Jungsun Oh, Yoonhee Kim, Jeeyoeng Kim, and Gwangyong Gim

Abstract South Korea is facing the issue of population decline alongside an increase in the elderly population. Changes in population structure have occurred due to a decrease in birth rates and an increase in the elderly population. The term "senior" typically refers to individuals aged 65 and above and is officially recognized in various countries and cultures. This study will use the term senior. The utilization of non-face-to-face methods, such as online education, remote work, online shopping, and telemedicine, has increased significantly due to the impact of the COVID-19 pandemic. As a result of these changes, there has been a growing desire among seniors to use smart devices for their social activities, convenience in daily life, hobbies, and health management. This study aims to empirically examine the factors influencing seniors' intention to use smart devices. It involved reviewing previous studies on seniors' characteristics and the characteristics of smart devices, followed by the design of a suitable model and hypothesis setting. The research model was based on the Technology Acceptance Model (TAM) and incorporated the concepts of innovativeness, self-efficacy, social influence, and convenience into perceived usefulness, along with the concept of absorption. The analysis results confirmed the reliability and validity of the variables, and it was validated that innovativeness, self-efficacy, social influence, convenience, perceived usefulness, and absorption influence seniors' intention to use smart devices. Lastly, this study suggests a policy and social direction that enhances seniors' satisfaction with the use of smart devices tailored to seniors' characteristics.

Keywords Senior · Smart devices · Technology acceptance model · Absorption

J. Oh (✉) · Y. Kim · J. Kim
Department of IT Policy and Management, Soongsil Univ., Seoul, South Korea
e-mail: jsohpost@gmail.com

Y. Kim
e-mail: kunice70@naver.com

G. Gim
Department of Business Administration, Soongsil Univ., Seoul, South Korea
e-mail: gygim@ssu.ac.kr

R. Lee et al. (eds.), *AI and Metaverse*, Studies in Computational Intelligence 1160,
https://doi.org/10.1007/978-3-031-61163-6_6

1 Introduction

In 2022, the senior population in South Korea aged 65 and above reached approximately 9,018,000 individuals, accounting for 17.5% of the total population, signifying the entrance into an aging society [1]. The senior population is rapidly increasing, necessitating product and service improvements tailored to senior characteristics.

Following the COVID-19 pandemic, societal and cultural changes, along with advancements in information technology, led to restrictions on social distancing and social interactions. Consequently, the utilization of smart devices increased significantly in fields such as remote education, remote work, and online shopping. These changes are driven by innovative smart devices based on technologies like big data, artificial intelligence, and the Internet of Things. There is growing interest in understanding how these changes impact seniors. An analysis of the factors that influence seniors' intention to use smart devices and subsequent research can contribute to improving seniors' quality of life and information accessibility. This, in turn, can offer value to relevant policies and service development.

2 Theoretical Background

2.1 Senior

Seniors are classified as an elderly population and are typically recognized as individuals aged 65 and above. They experience physical, cognitive, and social aging, which often leads to the need for assistance in their daily lives. The terms "senior," "elderly," "senior citizens," "older adults," and "silver generation" are commonly used to refer to this age group, but this study will use the term "seniors." Seniors are described as individuals who are in a phase of physiological and biological deterioration, experience changes in psychological aspects, including mental functioning and personality, and undergo alterations in their social status and roles [2].

Seniors have a growing interest in health, leisure activities, and social participation, and the government implements various welfare benefits and medical services to cater to their needs.

2.2 Smart Devices

Smart devices are devices that can perform various functions by connecting to the internet. Types of smart devices include smartphones, smart TVs, wearable devices, and they offer various apps and services, enabling functions such as video streaming, social media use, and online shopping. The innovative development of smart devices has brought about changes in both personal life and the utilization of corporate information systems. It has enabled the implementation and utilization of corporate information systems based on smart devices [3]. In other words, the innovativeness and convenience of smart devices are being extended to a wide range of services in enterprises, and this societal impact is influencing the intention to use smart devices.

This study focuses on researching the factors influencing seniors' intention to use smart devices.

2.3 Technology Acceptance Model (TAM)

The technology acceptance model states that a potential user's overall attitude toward the use of new technology is a major factor in determining usage of the new technology, and that attitude toward use is defined as a function of beliefs about perceived usefulness and perceived ease of use [4].

Perceived ease of use and perceived usefulness were used as mediators. These two variables influence attitudes, which, in turn, affect the intention to use, and the intention to use subsequently impacts actual system usage [5]. The technology acceptance model has been widely studied domestically and internationally to understand the relationship between behavioral intention and attitudes in the process of users adopting information technology. It is recognized as a model that explains the factors influencing the adoption of advanced products or technologies [6, 7].

This study utilized the technology acceptance model to analyze and understand the factors influencing the intention to use smart devices, based on this model.

2.4 Innovativeness

Innovativeness refers to the tendency to be an early adopter of new technologies, striving to become a pioneer in a specific technological domain, and demonstrating a leading attitude in embracing new technologies.

Prior research has substantiated the connection between innovativeness and the adoption of new technologies like smart devices [8]. It is known that the utilization of new technologies has a positive impact on activities such as video streaming, online shopping, online reservations, and more [9].

In this study, innovativeness is considered and adopted as a significant variable in seniors' intention to use smart devices, in response to societal changes.

2.5 Self-Efficacy

Self-efficacy is defined as an individual's judgment of their own abilities required to organize and execute actions to achieve a particular task [10]. It represents a subjective belief of a senior in their ability and confidence concerning the use of smart devices. People who believe they can easily handle new technology will experience higher self-efficacy, which, in turn, influences their attitude and behavior toward using that technology [11]. Previous studies have shown that self-efficacy influences perceived usefulness and the intention to use web browsing services, and self-efficacy, along with social influence, affects the intention to use [12].

Self-efficacy is a psychological characteristic widely applied as a variable in technology adoption models, which is why this study adopted self-efficacy as a variable in understanding seniors' intention to use smart devices.

2.6 Convenience

Convenience is a critical variable in technology adoption models, signifying the ease with which users can adopt and understand technology, ultimately enhancing user experience and efficiency. Research results have frequently shown that technologies perceived as convenient to use have a higher adoption rate [5]. In previous research, convenience was defined as the extent to which individuals perceive the use of smart devices in the context of in-house information systems as helpful for their tasks and for obtaining necessary information [3]. In this study, convenience related to assisting seniors in their daily lives through smart device use and the convenience of using smart devices were adopted as variables.

2.7 Social Influence

Social influence refers to the perceived pressure an individual receives to perform a specific behavior and is an important factor in product and technology adoption within the information technology field [9]. It is defined as the influence of social members on each other's behaviors [13]. According to prior studies, when seniors engage in positive social interactions and exert positive influence through social networking sites (SNS), they gain a sense of presence as network members through network effects [14].

In this study, social influence is adopted as a variable measuring the degree to which seniors believe and feel that people around them should use smart devices and the impact of this perception on their intention to use.

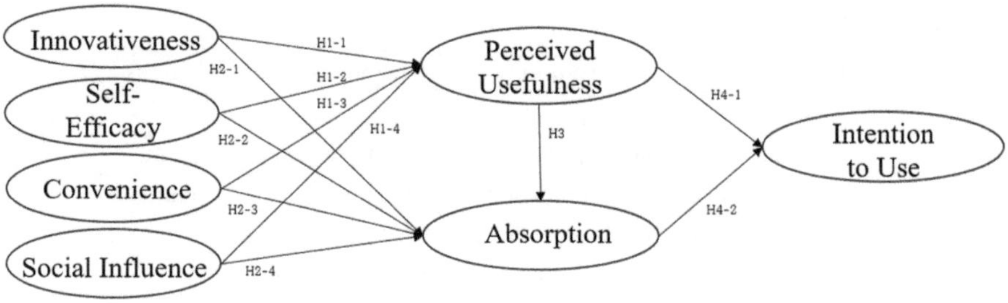

Fig. 1 Study model

2.8 Absorption

Absorption is explained as a personal predisposition or characteristic where a senior intensely focuses on a particular area of interest, leading to the overall attitude about a specific event [15]. Previous research conceptualized absorption as concentration, enjoyment, and time distortion, analyzed the impact of content characteristics on user absorption, and expanded the technology adoption model to study the impact of self-efficacy and social influence on smart device adoption [16]. In this study, the parameter of cognitive ease in the technology acceptance model was removed, and absorption was added to enhance explanatory power concerning seniors' intention to use smart devices.

3 Study Model and Hypotheses

3.1 Study Model

In this study, to validate the factors influencing seniors' intention to use smart devices, absorption was added based on the Technology Acceptance Model. This study aims to examine the factors that affect the intention to use based on the characteristics of smart devices and the characteristics of seniors, as extracted from prior research. The study model is shown in Fig. 1.

3.2 Hypotheses

Through a review of prior research, this study derived a total of 11 causal hypotheses related to the factors influencing seniors' intention to use smart devices. The hypotheses set in the study model are presented in Table 1.

Table 1 Hypotheses

H1–1	Innovativeness will have a positive (+) effect on perceived usefulness
H1–2	Self-efficacy will have a positive (+) effect on perceived usefulness
H1–3	Convenience will have a positive (+) effect on perceived usefulness
H1–4	Social Influence will have a positive (+) effect on perceived usefulness
H2–1	Innovativeness will have a positive (+) effect on absorption
H2–2	Self-efficacy will have a positive (+) effect on absorption
H2–3	Convenience will have a positive (+) effect on absorption
H2–4	Social Influence will have a positive (+) effect on absorption
H3	Perceived usefulness will have a positive (+) effect on absorption
H4–1	Perceived usefulness will have a positive (+) effect on the intention to use
H4–2	Absorption will have a positive (+) effect on the intention to use

3.3 Operational Definition of Variables

Based on the study model and hypotheses of this study, the operational definitions of variables and related research were defined. The definitions are shown in Table 2.

4 Research Methods and Empirical Analysis

4.1 Data Collection

For the data necessary for analysis, a questionnaire that can measure the study model was prepared based on prior research, followed by an online survey. The survey targets were seniors aged 65 or older, and the survey period was conducted for seven days from October 10, 2023, to October 16, 2023, and the contents of the survey were collected using a Likert 7-point scale and a total of 229 data were used for the final analysis.

4.2 Characteristics of the Sample

In terms of the demographic characteristics of the sample, as shown in Table 3.

Out of the total 229 respondents, there were 81 males and 148 females, accounting for a female proportion of 64.6%. Regarding the age distribution, the age group of 65–70 years old was the largest, comprising 57.6% of the respondents. Among the respondents, 216 individuals reported having prior experience in using smart devices.

Table 2 Operational Definition of Variables

Variable	Operational definition	Measurement item	Related research
Innovativeness	The tendency to use a new technology before others to become a pioneer in the field of new technology and the degree of attitude to lead in relation to the new technology	Use new technology before others The joy of using new technology Challenge the use of new technology A great deal of interest in new technology Efforts to acquire new skills	[17] [18] [3]
Self-efficacy	The degree of confidence and confidence in one's ability to use smart devices well	Can be used if someone else shows you how to use it Can be used when using similar experiences Can be used on user's manual Can be used if you see someone else using it Available for assistance from others	[19] [20]
Convenience	The degree to which you perceive that using smart devices is helpful to your life and can obtain the information you need	Easy compared to traditional systems Get the information you need easily Helpful in one's life Method to use is easy Easy to use	[21] [3]

(continued)

Table 2 (continued)

Variable	Operational definition	Measurement item	Related research
Social Influence	The degree of perception that people around me feel that I have to use a smart device	The influence of others The influence of people who are important to an individual People around me think it's useful to use Influential people think that I should use Most use it and recognize it as an advanced technology	[22] [9] [20] [3]
Perceived usefulness	The degree to which we believe smart device technology will improve the performance of users	Think it's useful to use The idea of relaxing everyday life The idea of making everyday life convenient The idea of being able to do everyday life efficiently The degree to which you can get daily life information quickly	[5] [20]

(continued)

Table 2 (continued)

Variable	Operational definition	Measurement item	Related research
Absorption	Degree of enjoyment and immersion by using smart devices	Degree of immersion in using smart devices How fun it is to use a smart device Degree of concentration of smart device use Degree of happiness in using smart devices Continuous use intention of using smart devices	[23] [17] [20] [16]
Intention to use	User tendency to decide whether to use the smart device and keep others informed in the future	Willingness to use in the future The need for in everyday life Positive propensity to use Frequency of use Recommend to others	[5] [24]

4.3 Factor Analysis and Reliability Analysis

In this study, the independent variables were defined as innovativeness, self-efficacy, convenience, and social influence. Exploratory factor analysis was conducted using SPSS 22 to validate the constructs. Principal component analysis was used for factor extraction, and Varimax rotation was employed.

In this study, all factors were considered significant as the factor loading values exceeded 0.5. After the factor analysis, Cronbach's alpha coefficients were utilized to assess the internal consistency reliability for each factor. The results of the analysis of Cronbach's alpha for each factor ranged from 0.926 to 0.950, indicating very high reliability, as shown in Table 4.

4.4 Regression Analysis and Hypothesis Testing

The results of multiple regression analysis examining the impact of the identified seniors' characteristic factors on usage intention are presented in Table 5.

Table 3 Population statistics

	Criterion	Frequency	Percentage
Gender	Male	81	35.4
	Female	148	64.6
	Total	229	100.0
Age	65–70	132	57.6
	71–75	50	21.8
	75–80	37	16.2
	81 or older	10	4.4
	Total	229	100.0
Usage	No	13	5.7
	Yes	216	94.3
	Total	229	100.0
The type of smart devices used	Smart TV	6	2.6
	Smart watch	43	18.8
	Smart phone	170	74.2
	Never used	10	4.4
	Total	229	100.0
Reason for selection	Brand	14	6.1
	Convenient to use	108	47.1
	Useful features	74	32.3
	Recommendation	23	10.1
	Never used	10	4.4
	Total	229	100.0

The analysis assessed the effects of innovativeness, social influence, convenience, and self-efficacy on seniors' intention to use smart devices. It also examined the mediating effect of perceived usefulness on the relationship between these factors and absorption. Finally, it investigated the impact of absorption on usage intention.

All 11 research hypotheses have been confirmed, verifying that all factors impact seniors' intent to use smart devices.

Table 4 Factor analysis and reliability analysis

Measurement items	Factor analysis				Reliability analysis
	Factor 1	Factor 2	Factor 3	Factor 4	Cronbach's alpha
Innovativeness 4	**0.854**	0.149	0.295	0.155	0.946
Innovativeness 5	**0.844**	0.100	0.239	0.245	
Innovativeness 2	**0.836**	0.207	0.192	0.268	
Innovativeness 1	**0.801**	0.174	0.182	0.269	
Innovativeness 3	**0.785**	0.211	0.263	0.267	
Social influence 3	0.176	**0.882**	0.169	0.133	0.926
Social influence 2	0.132	**0.857**	0.083	0.332	
Social influence 4	0.162	**0.820**	0.233	0.261	
Social influence 1	0.081	**0.764**	0.189	0.386	
Social influence 5	0.245	**0.701**	0.423	−0.111	
Convenience 4	0.308	0.184	**0.689**	0.403	0.926
Convenience 2	0.345	0.343	**0.682**	0.331	
Convenience 5	0.302	0.510	**0.664**	0.156	
Convenience 3	0.354	0.510	**0.662**	0.180	
Convenience 1	0.433	0.110	**0.606**	0.472	
Self-efficacy 3	0.402	0.324	0.211	**0.743**	0.950
Self-efficacy 4	0.456	0.334	0.340	**0.649**	
Self-efficacy 1	0.479	0.298	0.299	**0.637**	
Self-efficacy 2	0.460	0.298	0.301	**0.637**	
Self-efficacy 5	0.357	0.485	0.402	**0.517**	

Bold indicates the groups for each factor of measurement items

Table 5 Regression analysis and hypothesis testing

Dependent variable	Independent variable	R^2	B	S.E	β	t	p	Hypothesis	Acceptance status
Perceived usefulness	(Constant)	0.795	26.323	0.207		127.261	0.000		
	Innovativeness		2.099	0.207	0.307	10.123	0.000	H1–1	Accepted
	Social influence		4.176	0.207	0.610	20.143	0.000	H1–2	Accepted
	Convenience		3.890	0.207	0.568	18.767	0.000	H1–3	Accepted
	Self-efficacy		0.502	0.207	0.073	2.420	0.016	H1–4	Accepted
Absorption	(Constant)	0.674	22.716	0.282		127.261	0.000		
	Innovativeness		2.269	0.283	0.306	8.027	0.000	H2–1	Accepted
	Social influence		3.553	0.283	0.480	12.569	0.000	H2–2	Accepted
	Convenience		3.131	0.283	0.422	11.073	0.000	H2–3	Accepted
	Self-efficacy		3.070	0.283	0.414	10.860	0.000	H2–4	Accepted
Absorption	(Constant)	0.524	2.091	1.348		1.551	0.122		
	Perceived usefulness		0.784	0.050	0.724	15.804	0.000	H3	Accepted
Intention to use	(Constant)	0.859	−0.448	0.739		−0.607	0.545		
	Perceived usefulness		0.479	0.039	0.443	12.240	0.000	H4–1	Accepted
	Absorption		0.553	0.036	0.554	15.287	0.000	H4–2	Accepted

5 Conclusion

5.1 Conclusion and Implications

This study conducted empirical research on the factors influencing the intention to use smart devices among seniors. Previous research examined the characteristics of seniors and smart devices, and this study investigated them from the perspectives of perceived usefulness and absorption in smart devices. Previous research primarily focused on the intention to use smart devices across all age groups, highlighting the need for research based on user characteristics.

This study showed the following results.

First, Innovativeness, self-efficacy, convenience, and social influence significantly influence the intention to use smart devices through perceived usefulness and absorption. Social influence and convenience emerged as crucial factors affecting the intention to use smart devices among seniors. This suggests that seniors' use of smart devices is influenced by encouragement from family and friends, facilitating communication, and enhancing convenience in their daily lives. Secondly, seniors find enjoyment and happiness in smart devices, leading to greater absorption. Third, the intention to use smart devices among seniors is influenced more by the enjoyment and happiness they derive from using smart devices than by their perceived usefulness.

Lastly, this study emphasizes the significance of developing content that enhances absorption for seniors and creating user-friendly smart devices that align with the characteristics of seniors. Given the recent increase in the proportion of seniors, the findings of this study underscore the necessity of improving smart devices that cater to seniors.

This research highlights the importance of government policies and the establishment of an educational support system aimed at enhancing the quality of life and satisfaction of seniors through smart device usage.

5.2 Limitations and Future Research

This study has limitations because the study focused on a limited set of senior and smart device characteristics and was constrained by sample size and regional factors.

In the future, there is a need to further segment the characteristics of seniors and smart devices. Additionally, research should be conducted considering regional variations. Moreover, it is essential to expand research on the impact of smart device usage on the quality of life and satisfaction of seniors.

References

1. Statistics Korea (2022) 2022 Elderly statistics. Statistics Korea
2. Breen LB (1960) The aging individual. In: Tibbitts C (ed) Handbook of social gerontology. University of Chicago Press, Chicago
3. Koo EY (2015) A study on factors affecting the usage intention of smart devices-based information systems. Doctoral dissertation, Soongsil University
4. Davis FD (1985) A technology acceptance model for empirically testing new end-user information systems: theory and results. Doctoral dissertation, Massachusetts Institute of Technology
5. Davis FD (1989) Perceived usefulness, perceived ease of use, and user acceptance of information technology. MIS Quart 12:319–340
6. Kim GW (2016) A study on the factors influencing the internet of things (IoT) technology acceptance of SMEs: applying unified theory of acceptance and use of technology (UTAUT). Doctoral dissertation, Hansei University (2016)
7. Rogers EM (1995) Diffusion of innovation, 4th edn. Free Press, New York
8. Jeffers L, Atkin D (1996) Predicting use of technologies for communications and consumer needs. J Broadcast Electron Media 40(3):318–330
9. Venkatesh V, Davis FD (2000) A theoretical extension of the technology acceptance model four longitudinal field studies. Manag Sci 46:186–204
10. Bandura A (1977) Self-efficacy: toward a unifying theory of behavioral change. Psychol Rev 84(2):191–215
11. Davis FD, Bagozzi RP, Warshaw PR (1989) User acceptance of computer technology: a comparison of two theoretical models. Manag Sci 35(8):982–1003
12. Kim KK, Ryoo SY, Kim MO, Kim HJ (2009) Determinants of user intentions to use mobile web browsing service: self-efficacy and social influences. J Inform Technol Appl Manag 16(1):149–168
13. Venkatesh V, Brown SA (2001) A longitudinal investigation of personal computers in homes: adoption determinants and emerging challenges. MIS Quart 25(1):71–102
14. Kim JY, Chu KH (2014) The role of perceived value on the continuance intention in mobile social network service. J Dig Conver 12(10):211–222
15. Tellegen A, Atkinson G (1974) Openness to absorbing and self-altering experiences (" absorption"), a trait related to hypnotic susceptibility. J Abnorm Psychol 83(3):268
16. Lee SJ, Park JH, Kim JW (2010) Effects of content characteristics on the flow and perceived novelty: focused on characters and narratives of video UCC. J Korean Soc Des Sci 23(3):53–68
17. Agarwal R, Karahanna E (2000) Time flies when you're having fun: cognitive absorption and beliefs about information technology usage. MIS Quart 24(4):665–669
18. Kwon OJ (2008) An empirical study on the acceptance of the BSC system in the public sector. Doctoral dissertation, Dankook University
19. Venkatesh V, Morris MG, Davis GB, Davis FD (2003) User acceptance of information technology: toward a unified view. MIS Quart 27(3):425–478
20. Nam CH (2007) A study on factors influencing DMB adoption: focused on comparison between S-DMB and T-DMB. J Commun Sci 7(2):143–188
21. Kim JK, Nam ST (2012) Influence of smartphone characteristics on repurchase intention: considering moderating effect of playfulness. Korean J Bus Admin 25(4)
22. Taylor S, Todd PA (1995) Understanding information technology usage: a test of competing models. Inf Syst Res 6(2):144–176
23. Csikszentmihalyi M (1975) Beyond boredom and anxiety. Jossey-Bass, San Francisco
24. Bhattacherjee A (2001) Understanding information systems continuance: an expectation-confirmation model. MIS Quart 25(3):351–370

An Empirical Study on E-Government Service Using Web Standard Policy

Jeeyoeng Kim, Jungsun Oh, Gwangyong Gim, and Yoonhee Kim

Abstract The Republic of Korea's e-Government, represented by the Government24 (Gov.kr), is applying web standards to provide universal services to various devices and users. Recently, there has been an increasing effort to integrate South Korea's Big Tech with public services using AI and Cloud technologies. Gov.kr also provides services to South Korea's Big Tech through Open APIs. Furthermore, it aims to provide user-centered services that maximize user experience.

This paper presents the methods for applying service standardization and user-centered service provision that are being pursued by Gov.kr.

Keywords Government24 · Web policy · Open API · UI/UX

1 Introduction

In 2001, South Korea enacted the e-Government Act to establish the legal basis for the electronic processing of administrative tasks. Leveraging IT and information technology, it digitalized the operations of administrative and public institutions, constructing an e-Government to enable efficient performance of public services, including handling citizen complaints [1]. e-Government services aim to be accessible to all citizens and to achieve this goal, web standard policies are being implemented. Key web standard policies include web compatibility and web accessibility. Government24 (Gov.kr) is developing services in compliance with the standards

J. Kim (✉) · J. Oh · Y. Kim
Department of IT Policy and Management, Soongsil Univ., Seoul, South Korea
e-mail: kisulssa@gmail.com

Y. Kim
e-mail: kunice70@naver.com

G. Gim
Department of Business Administration, Soongsil Univ., Seoul, South Korea
e-mail: gygim@ssu.ac.kr

R. Lee et al. (eds.), *AI and Metaverse*, Studies in Computational Intelligence 1160,
https://doi.org/10.1007/978-3-031-61163-6_7

recommended by the World Wide Web Consortium (W3C). Moreover, with the emergence of new technologies such as AI, Big Data, and Cloud, and the growing use of API technologies, there is a trend towards service integration and convergence. Gov.kr aims to provide not only APIs but also user experience (UI/UX) designs that consider user convenience when accessing multiple e-Government services and plans to support a hassle-free unified login system (Any ID).

2 Background

2.1 *e-Government of Korea*

The e-Government in South Korea comprises two main areas: administrative business services utilized by public officials and citizen complaint processing services used by citizens or businesses. It encompasses various fields, such as public services/administration, disaster/safety, welfare/health, national tax/customs, and more. South Korea's e-Government has not only demonstrated excellence domestically but has also earned international recognition through assessments conducted by organizations like the UN and OECD, establishing itself as a leading e-Government nation in the global community. As a result, South Korea's e-Government has expanded its presence to 79 countries across North America/Central and South America, Europe, Africa, and Asia, and is being exported to various fields, including public safety and residents' daily lives. Projects like the 2023 Indonesia Cloud-Based Government Data Center Construction and the Mobile ID System Project in Malaysia are anticipated to be commissioned [2].

Government24, the integrated government services portal, is a representative e-Government service in South Korea that allows citizens to apply for or directly obtain complaint services online without visiting government offices in person. After applying for a complaint, citizens can directly check the processing results, ensuring full transparency throughout the process. Government24 does not store personal property, health, or personal information. To provide the services users require, Government24 is linked to around 300 public information systems operated by various ministries. Furthermore, Government24 offers one-stop services, which are services where systems from multiple ministries are interconnected to complete the process with a single application (Fig. 1).

2.2 *Web Standard Policy*

During the planning phase, before constructing e-Government services, standards and procedures for web standards that public agencies must adhere to are defined. Based on this, decisions on how to apply web standard technology to business plans

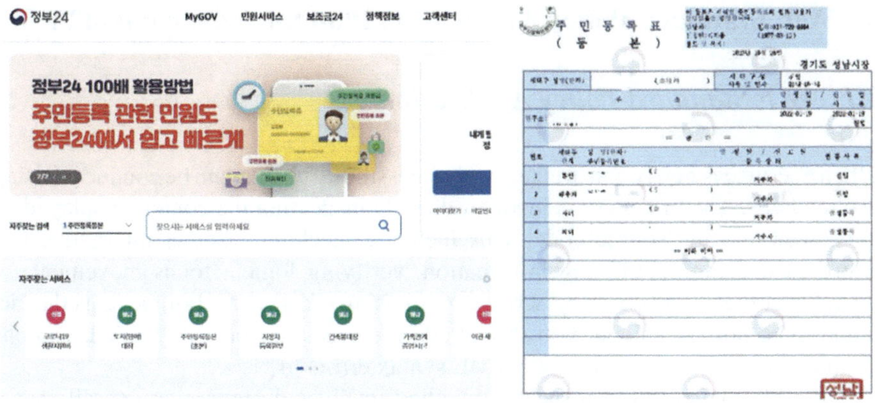

Fig. 1 Government24 website

are made when promoting information technology projects. The principle is to use standardized open technologies in the technology applied to information systems. Key considerations include applying standard technologies to ensure users can access services in various browser environments. Furthermore, design considerations should take into account people with disabilities, low-specification computer users, and other marginalized groups [3].

Web accessibility refers to the principle that when constructing or operating public information systems, individuals with disabilities, such as visual, auditory, or mobility impairments, as well as seniors, should be able to access websites or mobile applications without being hindered by their disabilities. To achieve this, the government has developed guidelines for website quality [4]. The key content of these guidelines emphasizes that regardless of disabilities, users should perceive all content provided on websites and app applications equally, and they should be able to operate all the functions provided by websites and app applications independently.

Web compatibility is a technique for creating web pages that use standardized web technology and do not optimize or depend on one side, such as operating systems and browsers, by using common elements. Problems arise when constructing web pages and mobile application content that are not compliant with web standards and are dependent on specific operating systems and browsers, making it impossible to use them on various operating systems and browsers. Additionally, the following should be adhered to in order to provide equal services even when the hardware and software environment of user terminals (PCs, mobile devices, etc.) is different [4]. First, the encoding method for characters should primarily use UTF-8. Second, the visual attributes of web pages (size, color, layout, alignment, and margins, etc.) should utilize international technical standards (CSS standards). Furthermore, functions for dynamically configuring and controlling web pages should use international technical standards (DOM standards).

3 Web Standard Policy Application Plan for Government24

3.1 Web Standard Policy Application

When a service is newly constructed, web accessibility checks can be conducted using two methods. The first method involves directly inspecting the content displayed on the screen and the source code. This includes checking whether alternative text is provided using HTML tag information, verifying logical focus movement for activated content, and ensuring scroll control using a keyboard. For any tags that are found to be non-compliant, you should add the necessary tags. If needed, you can use inspection tools to check for HTML syntax errors [4].

The web compatibility inspection method involves designers structurally separating the structure (HTML, XHTML) and presentation (CSS) to design the user interface (UI). Developers then implement the user interface according to the UI standard. The functionality of dynamically configuring and controlling web pages is checked to ensure it works equally in different browsers, and it is verified that JavaScript performs its intended functions correctly. Finally, it is essential to check whether web pages are displayed consistently across various browsers. This is typically evaluated using at least three different types of browsers, such as Chrome, Edge, Safari, etc., to determine if they provide equivalent functionality [4].

3.2 Continuous Application of Web Standard Policy

To enhance the development process for the application and maintenance of source code that adheres to web standards and accessibility, several measures are taken. Web publishers evaluate compliance with web standards and accessibility for service feature enhancements or new development projects. In cases where web standards and accessibility are not met, development source code can be transferred back to developers for necessary code improvements [5]. Web publishers also provide web standard guides to developers and assess code compliance with standards to ensure compatibility testing across various browsers and to secure accessibility. Regular training is used to encourage improved coding practices related to web standards among developers. Additionally, when collaborating with new or external resources for development, web standard guides and template codes should be provided to ensure adherence to web standards and accessibility [5].

4 Limitations and Improvement Strategies for Government24 Service

4.1 Limitations of Government24 Service

Government24, having been established in 2001, has consistently pursued expansion and construction projects over the course of approximately two decades. From a service perspective, Government24 was established to alleviate the inconvenience experienced by citizens who needed to individually verify and apply for services scattered across different government departments. However, the focus on quantitative expansion from the early stages of service development has led to a situation where the civil service provided by Government24 has expanded in quantity but lacks consistency and uniformity, ultimately causing inconvenience to users. Moreover, the presentation of various services on a single page has resulted in complex screen configurations. From a system perspective, Government24's services have been primarily centered around PCs, which has resulted in lower user satisfaction among those using smart devices. In addition, uncontrolled service development and expansion due to legal amendments and changing requirements have led to increased system complexity and a slowdown in service delivery.

4.2 Improvement Strategies for Government24 Service

Government24 has continuously strived to improve service quality, enhance user satisfaction, and expand the scope of online services. In 2023, it plans to further enhance services through user-centric support and service expansion. The scope of the project has been defined based on recent initiatives such as the "Digital Platform Government Implementation Plan" and "Government24 Information Strategy Plan (ISP)," as well as feedback from users requesting improvements to Government24 features.

Thus, this study aims to identify key areas and challenges that are crucial for the successful implementation of the Government24 enhancement project. To determine the critical issues in areas where quantitative analysis is challenging, the Analytic Hierarchy Process (AHP) method, which relies on qualitative knowledge, was employed. The process has been structured as follows (Fig. 2, Tables 1 and 2).

4.3 AHP Research Method and Analysis Results

AHP is a model for making decisions based on qualitative knowledge when quantitative analysis is challenging or partially quantitative [6].

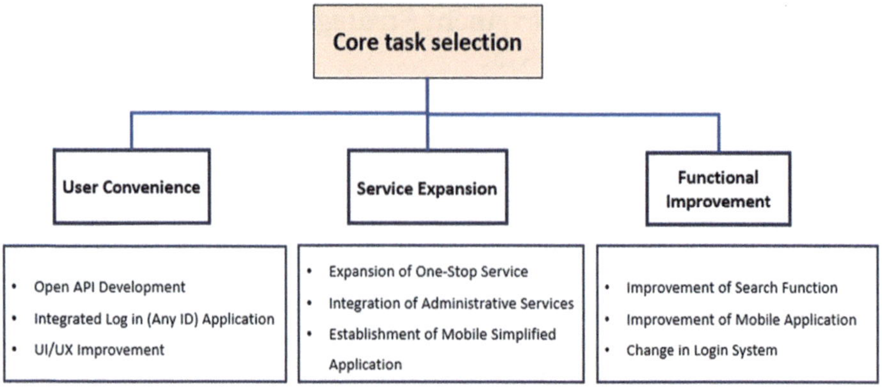

Fig. 2 Hierarchical model

Table 1 Hierarchical model explanation (top-level categories)

Top-level categories	Main contents
User convenience	– Area of service quality improvement from the perspective of Government24 users
Service Expansion	– Area of expanding online government services in line with the growth of non-face-to-face services
Functional improvement	– Area of improving errors and complex transaction procedures that occur during service use

To ensure the reliability of the research model, factors were derived based on South Korea's 'Digital Platform Government Realization Plan,' 'Government24 Information Strategy Plan,' and 'Improvement Requests from Government24 Users.' The survey respondents were divided into two groups: public-sector experts and private-sector experts. Public sector experts were selected from those responsible for planning and designing national informatization strategies, planning and constructing electronic government services, and evaluating national informatization projects, including professors in charge of the return system. Private sector experts were selected based on their expertise in constructing the Government24 system or providing consulting on future development directions.

Thus, a survey was conducted with seven public and private sector experts, who constituted the survey participant group. The purpose of the research, the survey method, and the AHP model were explained in advance, and pairwise comparison surveys were conducted. The AHP questionnaire was developed using the commonly applied 9-point scale, and the survey results and summaries were analyzed using the Super Decision tool.

[Analysis Results for Top-Level Categories] "User Convenience" was selected as the priority category for consideration in the Government24 Advanced Project in 2023. "Service Expansion," which mainly appeared in the early stages of service

Table 2 Hierarchical Model Explanation (Subdivisions)

Tope-level categories	Subdivisions	Key contents
User convenience	Open API development	– Utilization of the Government24 service on private platforms – Issuance through private portals without accessing Government 24
	Integrated log-in (Any ID) application	– Absence of an integrated electronic government service authentication system – Unified Provision of Government Services using a Single Account
	UI/UX improvement	– Provision of inconsistent UIs for different government services – Provision of common UIs and maximization of user experience
Service expansion	Expansion of one-stop service	– Integration of fragmented service information across ministries – Batch processing with a single application in Government24
	Integration of administrative services	– Increased demand for contactless services due to COVID-19 pandemic – Expansion of services that can be processed online
	Establishment of mobile simplified application	– Identification and analysis of services frequently applied by citizens in Government24 – Service development to enable applications through mobile devices
Functional improvement	Improvement of search function	– Provision of personalized services by linking login information – Provision of recommended services linked to user history information
	Improvement of mobile application	– Optimization of content placement through user process analysis – Simplification of member registration and implementation of adaptive civic application services
	Change in login system	– Restructuring of the Government24 login authentication system – Simplification of input procedures for biometric information (fingerprints) registration

construction, was found to be the lowest in priority, indicating that Government24 has moved from its initial stage to a mature stage. The analysis results are explained in the order of priority, with "User Convenience" as the first category, followed by "Function Improvement," and "Service Expansion" as the last (Table3).

[Subdivision Analysis Results by Sector]

① In the 'User Convenience' category, there are three specific items. All three items contribute to providing users with the maximum convenience. In particular, the top-priority selection of 'Open API Development' signifies the standardization of processes within Government24 and the utilization of the developed outcomes on private platforms. The application of Any-ID and UI/UX improvement aims to address the inconveniences associated with user account management and the lack of consistency in screen layout and menu structure experienced by users of various government websites.

② In the 'Functional Improvement' category, the highest priority was assigned to the reformation of the Government24 login system, which significantly outweighed the other items, such as the enhancement of the search function and mobile application. This suggests that there are numerous inconveniences in the login methods, processes, and procedures that users must go through when using Government24. Results from online satisfaction surveys also reflected a substantial number of improvement requests in this area. Given that Government24 provides services that include personal information, property information, and health information, it is presumed that these issues are a result of the need for enhanced identity verification processes.

③ In the 'Service Expansion' category, Expansion of One-Stop Service ranked second among the sub-items. This indicates that, while improving 'user convenience' is essential, there is a high priority for providing services that allow citizens

Table 3 Analysis result

Top-level categories	Weight	Priority	Subdivisions	Weight	Priority
User convenience	0.467	1	Open API development	0.175	1
			Any ID application	0.148	3
			UI/UX improvement	0.144	4
Service expansion	0.259	3	Expansion of one-stop service	0.157	2
			Integration of administrative services	0.056	8
			Establishment of mobile simplified application	0.047	9
Functional improvement	0.274	2	Improvement of search function	0.063	7
			Improvement of mobile application	0.084	6
			Change in login system	0.126	5

to handle complex administrative tasks with a single click. One-stop service is favored for its ability to streamline a wide range of administrative procedures, making it convenient for various segments of the population, such as working professionals, seniors, and housewives. Additionally, the integration of administrative services is expected to expand due to the increased demand for online, contactless services brought about by the COVID-19 pandemic. It is planned to extend the range of services that can be applied online through Government24 by enhancing interoperability between different government agency systems.

Based on the survey results from experts in various fields, key areas to consider when constructing Government24 services were identified, and high-priority sub-items (tasks) were selected. This paper will explain the approach of applying Open API and UI/UX design standards related to web standards to Government24, as derived from the AHP results in the core areas. Additionally, this paper will explain the approach to implementing an integrated login system (Any ID) that enables users to access a variety of government e-services under one account, although it's not a web standard technology per se.

5 Government24 Service Application Plan

5.1 Open API Application Plan

Government24 aims to enhance accessibility to government services by opening up and promoting the integration of private and public services, as well as generating a broad societal impact. To achieve this goal, Government24 will develop widely used public services as standard APIs and register them on a government-operated open platform. This will enable businesses to utilize the registered API services on their platforms, allowing Government24 users to access services not only through the Government24 website or mobile application but also through their preferred private sector platforms. To implement Open APIs, the specific steps are as follows. Government24 will identify the most frequently used services and develop API services for integration with private sector platforms. These developed API services will be registered on a cross-government service open platform, making them available for use on various business platforms following approval (Fig. 3).

5.2 UI/UX Application Plan

An intuitive and consistent UI/UX design to make public websites and mobile applications easily accessible and user-friendly for everyone will be implemented. The plan is to develop standard models for different user types based on the analysis of user behavior patterns and user histories, among other data. These user-type specific

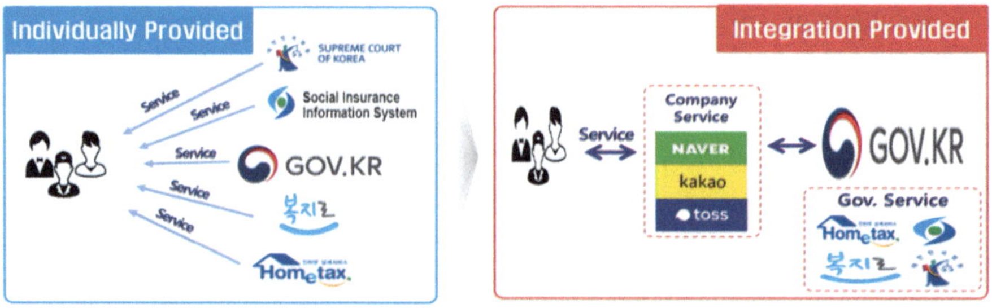

Fig. 3 Public service open API

standard models take into account metrics such as click rates, time spent on the platform, and bounce rates, allowing the optimization of the user experience for each user category through various simulations [7]. Additionally, enhancement of the overall application process by improving how information is presented to users at each processing stage when they apply for civil services will be made. This includes displaying relevant information to users based on the specific stage they are at and providing a more visually informative representation of the progress.

5.3 *Integrated Log-In (Any ID) Application Plan*

Most users access Government24 through corporate portal sites such as Google and Naver, and the primary source of inconvenience reported during the use of Government 24 services pertains to the registration and login processes. As an alternative solution, various authentication methods have been requested [8]. Thus, Government24 is establishing a unified login system, allowing users to log in to all public websites and mobile applications using the authentication method of their choice, without the need to create separate IDs for each government site. To achieve this, Government24 will implement a single-screen display of various login methods that users prefer. Furthermore, Government24 will enable the sharing of user information across government services, such as Home Tax, Welfare Services, Employment 24, and more, through Single Sign-On (SSO) linkage following login. Additionally, differentiating services based on authentication login methods, Government24 will assign distinct security levels to services requiring identity verification [8] (Fig. 4).

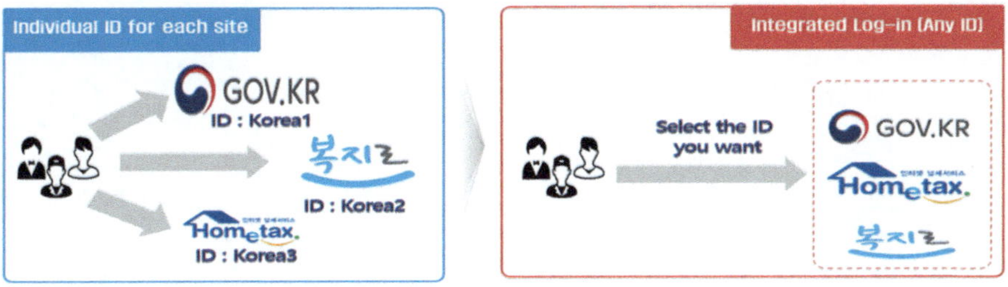

Fig. 4 Integrated log-in (any ID) application method

6 Conclusion

Since 2000, South Korea has been providing online services to its citizens and enhancing the productivity of public servants through the e-Government Roadmap projects, delivering services needed by the public. South Korea is transitioning beyond e-government and digital government to a Digital Platform Government. The Digital Platform Government integrates all data on a single platform, allowing citizens, businesses, and the government to collaborate in solving societal issues and creating new value [7]. The ultimate goal of the Digital Platform Government is to connect all data through a single platform and leverage artificial intelligence, big data, cloud technologies, and more to become the world's leading government.

At the core of the Digital Platform Government is Government24. Government24 is pursuing an Information Strategy Plan (ISP) to achieve "One Government, One Site." Through this project, Government24 is designing the vision and target model for the next-generation Government24, along with formulating a strategic plan [8]. The target model selected through the ISP will be progressively implemented starting from 2024.

References

1. Electronic Government Act. https://www.law.go.kr/
2. Korea IT News, E-government export exceeded $500 million... "Brighter future this year". https://english.etnews.com/20230126200006
3. Guidelines for Establishment and Operation of Information System for Administrative and Public Institutions. https://www.law.go.kr/
4. Guidelines for Quality Control on e-Government Websites. https://www.law.go.kr/
5. Kim JY, Lee SH, Choi JY (2010) A system management plan for enhancing web standard and accessibility. In: Proceedings of the 37th KIISE Fall Conference, pp 160–163
6. Hong DP (2023) A study on prioritization and importance analysis of hospital advertising items using Delphi and AHP methods. (Doctoral Dissertation) Hanyang University, pp 17–18
7. MOIS (2022) The digital platform government promotion plan of the ministry of the interior and safety
8. MOIS (2023) BPR/ISP completion report for integrated platform for administrative services

The Effects of Surface and Deep Acting on Childcare Teacher's Turnover Intention: Mediating Effect of Organizational Commitment and Moderating Effect of Leader-Member Exchange (LMX)

Sung-Hee Baek, Ho-Chul Shin, Mi-Jeong Kim, and Mi-Sook Han

Abstract This study aims to explore the effect of childcare teachers' emotional labor strategies (e.g., surfacing acting or deep acting) on their Turnover Intention, a significant factor for preventing turnover and enhancing the quality of childcare services. In addition, the study seeks to the effect of organizational commitment as a mediator in the effect of emotional labor strategy on turnover intention. The results of the study are as follows: First, surface acting among the emotional labor strategies had a positive (+) effect on the Turnover Intention, while deep acting had a negative (−) effect. Second, organizational commitment mediated the relationship between both emotional labor strategies and the Turnover Intention. Finally, the leader-member exchange acted as a moderating variable, weakening the relationship between deep acting and turnover intentions.

Keywords Emotional labor strategy · Turnover intentions · Organizational commitment · Leader-member exchange (LMX)

S.-H. Baek · M.-J. Kim · M.-S. Han
Department of Project Management, Graduate School of Soongsil University, Seoul, Korea
e-mail: b0s0h073@naver.com

M.-J. Kim
e-mail: perfumejung@naver.com

M.-S. Han
e-mail: sookhan01@naver.com

H.-C. Shin (✉)
School of Business Administration, Soongsil University, Seoul, Korea
e-mail: hshin@ssu.ac.kr

R. Lee et al. (eds.), *AI and Metaverse*, Studies in Computational Intelligence 1160,
https://doi.org/10.1007/978-3-031-61163-6_8

1 Introduction

The modern society, characterized by a decrease in family members and an increase in dual-income couples, has seen a decline in the educational functions of the family, emphasizing the role of childcare teachers as educators of young children [1]. However, despite their importance, childcare teachers are being called upon for ceaseless love and devotion for infants and increased emotional commitment [2].

This has led to a surge in stress due to low social recognition, heavy workload, low wages, physical and emotional fatigue, a sense of powerlessness, and psychological burden, which are associated with a low level of personal achievement, leading to job changes [3]. In relation to the high turnover rate of childcare teachers, there is a growing interest in emotion labor strategies in the field of childcare. Emotion labor strategies are behaviors that express emotions suitable for emotion expression rules required by an organization [4], which frequently occur in high emotion labor situations among childcare teachers working in kindergartens.

According to previous studies on emotion labor strategies, there have been reports that the turnover intention is caused by the emotion labor strategies of childcare teachers in kindergartens, as the necessity of effective emotion management of childcare teachers has been raised [5]. If emotion labor strategies act as important predictive factors for turnover intention, it is believed that research is needed to clarify the process of how emotion labor strategies of childcare teachers affect turnover intention and to identify a third variable to reduce turnover intention.

In this study, we focused on whether organizational commitment plays a mediating role when emotion labor strategies influence turnover intentions. Organizational commitment is considered a major factor from the organizational aspect among various turnover decision factors, and it is supported that it is related to the decrease in the turnover rate of childcare teachers [6], so it can be evaluated as a representative factor that performs a positive function for achieving organizational goals in the subjective environment of organizational behavior. Furthermore, we aim to empirically clarify the mechanism through which emotion labor strategies and turnover intention influence each other, not only by examining the relationship between emotion labor strategies and turnover intention, but also by using organizational commitment as a mediator between emotion labor strategies and turnover intention.

2 Theoretical Background

2.1 *Emotional Labor Strategies*

Emotional labor strategies refer to the control and expression of one's emotions through facial expressions or physical gestures, in a way that can be observed by others, in order to conform to the organization's rules of emotional expression for job performance [7]. There are two core techniques of emotional labor strategies: surface acting and deep acting [4]. Surface acting involves merely controlling one's emotional expressions outwardly without changing the actual emotions, a passive engagement with emotional regulation, where the true feelings do not align with the emotional expression rules directed by the organization. On the other hand, deep acting is an act of active involvement in emotional control, which means changing the internal emotions of members of the organization in order to evoke desired emotions [8].

2.2 Turnover Intention

The Turnover Intention can be described as a predictive variable for identifying turnover, which is defined as the intention to depart from the current job [9]. While turnover intention and actual turnover are not identically defined, analyzing the variables that influence turnover intention could play a preventative role against actual turnover. The theoretical models explaining turnover intention include Expectancy Disconfirmation Theory, Social Exchange Theory, and Organizational Socialization Theory.

2.3 Organizational Commitment

Generally, commitment implies a combination of an individual's attitude and the will to act, signifying the possession of an identity towards a particular entity or contributing to it [10]. This sense of unity and belonging can be seen as the basic concept of organizational commitment for an employee towards their organization. This study classified organizational commitment not just as a single-dimensional factor but as consisting of three components: affective commitment, continuance commitment, and normative commitment. Previous research has shown that higher affective commitment positively influences the organization, whereas continuance commitment, characterized by calculative and exchange aspects, may not be seen as only a positive factor [11]. Moreover, normative commitment relates to a sense of duty or responsibility to act towards the achievement of organizational goals, indicating that affective commitment is the state of wanting to remain in the organization, continuance commitment arises from having to stay, and normative commitment stems from the sense of ought to stay [12].

2.4 Leader-Member Exchange (LMX)

The Leader-Member Exchange relationship explains the interdependent relationship between leaders and organizational members. The LMX relationship has been a subject of interest to researchers for decades in providing a deeper understanding of how leaders form productive relationships with members to achieve organizational goals [13]. Research on LMX theory focuses on the impact of relationships that extend beyond the employment contract between leader and member and is based on theories like Social Exchange Theory and Role Marking Theory, emphasizing that higher levels of LMX result in more positive emotions and trust and more stable and positive social exchange relations through frequent LMX interactions [14]. In sum, LMX theory focuses on the dyadic relationships between leaders and members, suggesting that leaders do not influence all members in a uniform way but develop different types of relationships with each member. This study seeks to empirically discuss the components of LMX, defined as emotional bonds, loyalty, contribution, and respect for professionalism.

3 Research Method

3.1 Research Model and Research Hypothesis

This study reviewed the concepts defined in previous studies and the relationships between the variables, revised and designed the research model, and proposed the following research model and research hypotheses (Fig. 1).

According to [15], emotional labor strategies are defined as the surface acting that an organization member engages in when they express emotions required by the

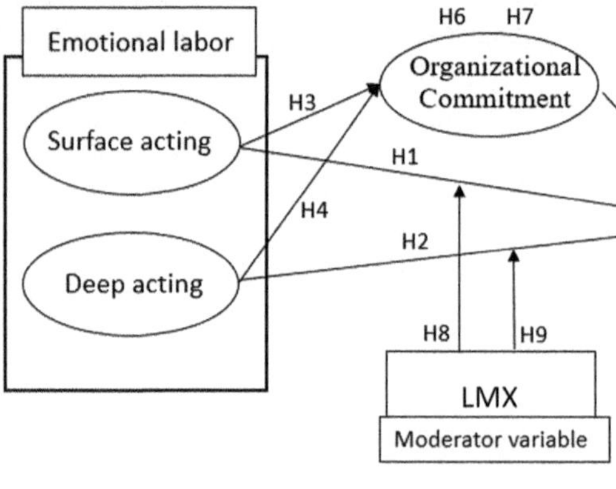

Fig. 1 Research Models

organization, which differ from their genuinely felt emotions. Surface acting includes behaviors that are coerced due to organizational demands, entailing a divergence from the actual emotions, thus encompassing the process of suppressing the real emotions. This can lead to negative psychological states [8]. Moreover, surface acting, associated with inauthentic emotions, turnover intentions, and emotional exhaustion, suggests that a higher level of surface acting could positively influence the Turnover Intention [8]. Conversely, deep acting refers to when an individual actually feels or experiences the emotion they are attempting to convey in alignment with the organization's display rules [7]. Deep acting involves altering one's actual emotions to match the emotions expressed, meaning that organization members transform their emotions into those deemed desirable by the organization [8]. Consequently, when organization members attempt to genuinely feel the emotions they express, this can foster a positive psychological state and lead to positive expressions [16]. Empirical studies on emotional labor strategies have revealed that positive emotional experiences of organization members are associated with turnover intentions, asserting that positive emotions correlate negatively with the Turnover Intention [17]. Based on the foregoing literature, we propose the following hypotheses concerning the relationship between emotional labor strategies, divided into surface and deep acting, and the Turnover Intention.

Hypothesis 1. Surface acting will have a positive (+) influence on the Turnover Intention.

Hypothesis 2. Deep acting will have a negative (−) influence on the Turnover Intention.

Organizational commitment is known to positively impact the performance of organization members and decrease their Turnover Intention, thus benefiting the productivity of society at large as well as individuals and organizations [13]. Research on organizational commitment from an emotional standpoint emphasizes that it involves members sharing organizational values and perceiving themselves as part of the organization [14]. The study of organizational commitment has significant implications for emotional labor strategies. Many scholars have viewed surface acting as detrimental to organizational commitment due to emotional dissonance. Morris and Feldman [18] confirmed that surface acting as an emotional labor strategy negatively impacts organizational commitment, and [16] also supported this finding, arguing that surface acting significantly affects organizational commitment negatively. Additionally, recent studies report that as deep acting increases, so does organizational commitment among members. Notably, numerous studies investigating the relationship between emotional labor strategies and organizational commitment have shown significant positive impacts of deep acting on organizational commitment [18]. Based on these results, the following hypotheses are presented for the relationship between emotional labor strategies, segmented into surface and deep acting, and organizational commitment.

Hypothesis 3. Surface acting will have a negative (−) influence on organizational commitment.

Hypothesis 4. Deep acting will have a positive (+) influence on organizational commitment.

Organizational commitment, indicating the loyalty and attachment of organization members to their organization, has been supported by numerous studies as having a negative impact on the Turnover Intention [14]. Allen and Meyer [14] reported that organizational commitment could relatively explain turnover rates and organizational performance, indicating that higher organizational commitment correlates with better performance. Additionally, the operational definition of organizational commitment in previous studies suggests that higher levels of commitment are indicative of a lower Turnover Intention [18]. This research aims to verify the direct effects of organizational commitment on the Turnover Intention and to compare the relative influence of emotional labor strategies and the Turnover Intention.

Hypothesis 5. Organizational commitment will have a negative (−) influence on the intention to leave.

Homans [11] argued that surface acting, which requires suppressing genuine emotions to express those demanded by the organization, hinders the emotional autonomy of organization members, potentially lowering organizational commitment and increasing the Turnover Intention. Research by Stazyk et al. [19] found a partial mediating effect of organizational commitment between surface acting and the Turnover Intention. It was also reported that surface acting positively influences the Turnover Intention and negatively affects organizational commitment. These previous findings allow the prediction of a research hypothesis that organizational commitment will mediate the relationship between surface acting and the Turnover Intention. Empirical studies on emotional labor strategies have shown that positive emotional experiences of organization members are related to the Turnover Intention [8, 17] argued that positive emotions have a negative (−) relationship with the Turnover Intention. Building on this literature, the current study sets forth the following hypotheses to examine the mediating effects of organizational commitment on the relationship between emotional labor strategies and the Turnover Intention.

Hypothesis 6. Organizational commitment will mediate the relationship between surface acting and the Turnover Intention.

Hypothesis 7. Organizational commitment will mediate the relationship between deep acting and the Turnover Intention.

The survey was conducted over a 55-day period, from June 6 to July 31, 2022, targeting 350 childcare center staff members in Gyeonggi Province, resulting in 334 returned questionnaires. After excluding 26 insincere responses, the final analysis utilized 311 questionnaires. For empirical analysis of the proposed research model, data were collected via Google Forms from childcare teachers and verified statistically. The collected data were analyzed using the SPSS 28 software following several steps. Firstly, demographic analysis of the participant childcare teachers was conducted through frequency analysis. Secondly, exploratory factor analysis was performed to validate the research variables, followed by the computation of Cronbach's α values for reliability verification. Pearson's correlation coefficients were then calculated to understand the relationships among surface acting, deep acting, organizational commitment, Leader-Member Exchange, and Turnover Intention. Finally, hierarchical regression analysis was conducted to discern the causal relationships between surface and deep acting, organizational commitment, Turnover Intention, and the mediating effects of the Leader-Member Exchange relationship.

Hypothesis 8. The relationship between surface acting and Turnover Intention will be weakened by LMX.

Hypothesis 9. The relationship between deep acting and Turnover Intention will be strengthened by LMX.

3.2 Data Collection and Analysis Methods

The survey The survey was conducted on 350 childcare teachers located in Gyeonggi-do from June 6 to July 31, 2022, and 334 copies of the questionnaire were collected. For the collected data, demographic analysis of participating childcare teachers was conducted through frequency analysis using SPSS 28 software Data Collection and Analysis Methods.

The survey was self-administered and comprised 52 items in total: 13 items for emotional labor strategies, 24 for organizational commitment, 12 for Leader-Member Exchange (LMX), and 3 for Turnover Intention. In addition to these, 10 demographic items were included to profile the respondents, making up a total of 62 items used in the survey. Emotional labor strategies, Turnover Intention, organizational commitment, and leader-member exchange were all measured using a 5-point Likert scale, excluding the statistical demographic item.

4 Data Analysis

4.1 Verification of Research Hypotheses

4.1.1 The Relationship Between Emotional Labor Strategy (Surface Behavior, Inner Behavior) and Turnover Intention

Table 1 shows the results of verifying Hypothesis 1, which predicts that the surface behavior of childcare teachers will have a positive (+) effect on turnover intention, and Table 2 shows the results of verifying Hypothesis 2, which predicts that the inner behavior of childcare teachers will have a negative (−) effect on turnover intention.

4.1.2 The Relationship Between Emotional Labor Strategy (Surface Behavior, Inner Behavior) and Organizational Commitment

Table 3 shows the results of verifying Hypothesis 3, which predicts that the surface behavior of childcare teachers will have a negative (−) effect on turnover intention,

Table 1 The effect of surface behavior on turnover intention

Varlable		Turnover intention			
		Model 1		Model 2	
		ß	t	ß	t
(Aconstantnumber)			10.613***		4.974***
	Age in charge	−0.059	−1.027	−0.055	−0.978
Independent	Surface behavior			0.201	3.541***
R^2		0.055		0.092	
ΔR^2		0.055		0.037***	
F		3.354**		5.164***	

*p < 0.05, **p < 0.01, ***p < 0.001

Table 2 The effect of inner behavior on turnover intention

Varlable		Turnover intention			
		Model 1		Model 2	
		ß	t	ß	t
(Aconstantnumber)			10.613***		12.632***
Independent	Inner behavior			−0.360	−6.690***
R^2		0.055		0.176	
ΔR^2		0.055		0.121***	
F		3.354**		10.845***	

*p < 0.05, **p < 0.01, ***p < 0.001

and Table 4 shows the results of verifying Hypothesis 4, which predicts that the inner behavior of childcare teachers will have a positive (+) effect on organizational commitment.

Table 3 The effect of surface behavior on Organizational commitment

Varlable		Organizational commitment			
		Model 1		Model 2	
		ß	t	ß	t
(Aconstantnumber)			12.193***		13.448***
Independent	Surface behavior			−0.337	−6.209***
R^2		0.064		0.170	
ΔR^2		0.064		0.106***	
F		4.201**		10.357***	

*p < 0.05, **p < 0.01, ***p < 0.001

Table 4 The effect of inner behavior on Organizational commitment

Varlable		Turnover intention			
		Model 1		Model 2	
		ß	t	ß	t
(Aconstantnumber)			12.193^{***}		4.618^{***}
Independent	Organizational commitment				
R^2				0.306	5.608^{***}
ΔR^2		0.064		0.135	
F		0.064		0.089^{***}	
		4.201^{**}		9.902^{***}	

*p < 0.05, **p < 0.01, ***p < 0.001

Table 5 The effect of Organizational commitment on turnover intention

Varlable		Organizational commitment			
		Model 1		Model 2	
		ß	t	ß	t
(Aconstantnumber)			10.613^{***}		15.212^{***}
Independent	Organizational commitment			−0.475	-9.352^{***}
R^2		0.055		0.266	
ΔR^2		0.055		0.211^{***}	
F		3.354^{**}		18.378^{***}	

*p < 0.05, **p < 0.01, ***p < 0.001

4.1.3 The Relationship Between Organizational Commitment and Turnover Intention

Table 5 shows the results of regression analysis to verify Hypothesis 5, which predicted that the organizational commitment of childcare teachers would have a negative (−) effect on turnover intention.

4.1.4 Mediating Effect of Organizational Commitment

Table 6 shows the results of verifying Hypothesis 6, which predicts that childcare teachers' organizational commitment will mediate the relationship between surface behavior and turnover intention, and Table 7 shows the verification results of Hypothesis 7, which predicts that childcare teachers' organizational commitment will mediate the relationship between inner behavior and turnover intention.

Table 6 The mediating effect of Organizational commitment in the relationship between surface behavior and turnover intention

Varlable		Organizational commitment					
		Model 1		Model 2		Model 3	
		ß	t	ß	t	ß	t
(Aconstantnumber)			10.613***		4.974***		9.582***
Independent	Surface behavior			0.201	3.541***	0.046	0.850
Parameters	Organizational commitment					−0.460	−8.521***
R^2		0.055		0.092		0.268	
ΔR^2		0.055		0.037***		0.176***	
F		3.354**		5.164***		15.841***	

*p < 0.05, **p < 0.01, ***p < 0.001

Table 7 The mediating effect of Organizational commitment in the relationship between inner behavior and turnover intentiont

Varlable		Turnover intention					
		Model 1		Model 2		Model 3	
		ß	t	ß	t		
(Aconstantnumber)			10.613***		12.632***		15.349***
Independent	Surface behavior			−0.360	−6.690***	−0.236	−4.580***
Parameters	Organizational commitment					−0.403	−7.788***
R^2		0.055		0.176		0.314	
ΔR^2				0.037***		0.137***	
F		3.354**		10.845***		15.841***	

*p < 0.05, **p < 0.01, ***p < 0.001

5 Conclusion

In this study, the importance of organizational commitment as a mechanism for positive outcomes, managed through the relationship between leaders and organization members, and as a path through which the sub-factors of emotional labor strategies, such as surface acting and deep acting, influence childcare teachers' Turnover Intention was confirmed. This suggests the significance of a horizontal mutual trust and respect relationship between leaders and organization members, and organizational commitment to stabilize childcare teachers' performance and reduce negative factors in overall childcare service quality, thereby lowering their Turnover Intention. These are vital for promoting the desirable emotional and behavioral development of young children and the overall quality of childcare services.

The empirical analysis of this study did not show the effect of Leader-Member Exchange (LMX) on the relationship between surface acting and the Turnover Intention. This implies that even if childcare teachers at daycare centers try to understand infants and parents, the emotional labor strategy inherent to the nature of their duties may potentially lead them to contemplate leaving. However, the empirical analysis revealed that high levels of deep acting and LMX correlate negatively with the Turnover Intention, which is due to psychological rewards such as loyalty and recognition that foster a positive attitude in childcare teachers before considering leaving.

Therefore, daycare centers should strive to establish stable childcare environments and positive teacher retention through active exchange relationships between leaders and organization members.

References

1. Coplan RJ, Bullock A, Archbell KA, Bosacki S (2015) Preschool teacher's attitudes, beliefs, and emotional reactions to young children's peer group behaviors. Early Childhood Res 30(1):17–27
2. Zhang Q, Zhu W (2008) Exploring emotion in teaching: emotional labor, burnout, and satisfaction in Chinese higher education. Commun Educ 57(1):105–122
3. Seefeldt C, Denton K, Galper A, Younoszai T (1998) Characteristics of former headstart parents, awareness of school atmosphere, participation in children's education. Element J 98(4):339–349
4. Ashforth BE, Humphrey RH (1993) Emotional labor in service roles: the influence of identity. Acad Manag Rev 18(1):88–115
5. Zembylas M (2005) Beyond teacher cognition and teacher beliefs: the value of the thnography of emotions in teaching. Int J Qual Stud Educ 18(4):465–487
6. Bluedorn AC (1982) A unified model of turnover from organizations. Hum Relat 35(2):135–153
7. Hochschild A (1983) Comment on Kemper's social constructionist and positivist approaches to the sociology of emotions. Am J Sociol 89(2):432–434
8. Grandey AA (2003) When the show must go on surface acting and deep acting as determinants of emotional exhaustion and peer-rated service delivery. Acad Manag J 46(1):86–96
9. Iverson RD (1992) Employee intent to stay: an empirical test of a revision of the price and Mueller model. The University of Iowa, Ann Arbor, pp 50–65
10. Steers RM, Mowday RT (1981) Employee turnover and post-decision justification. Res Org Behav 3(1):235–282
11. Homans GC (1961) The humanities and the social sciences. Am Behav Sci 4(8):3–6
12. Lutfey K, Mortimer JT (2003) Development and socialization through the adult life course. Handbook of social psychology. Springer, Boston, MA, pp 183–202
13. O'Reilly C, Chatman J (1986) Organizational commitment and psychological attachment: the effects of compliance, identification and internalization on prosocial behavior. J Appl Psychol 71(3):492–499
14. Allen NJ, Meyer JP (1993) Organizaitonal commitment: evidence of career stage effects. J Bus Res 26(5):49–61
15. Schriesheim CA, Castro SL, Coglister CC (1999) Leader-member exchange (LMX) research: a comprehensive review of theory, measurement and data-analytic practices. Leadership Quart 10(1):63–113

16. Brotheridge CM, Lee RT (2003) Development and validation of the emotional labour scale. J Occup Organ Psychol 76(3):365–379
17. George JM, Jones GR (1996) The Experience of work and turnover intentions: inter active effects of value attainment, job satisfaction, and positive mood. J Appl Psychol 81(3):318–359
18. Morris JA, Feldman DC (1996) The dimensions, antecedents, and consequences of emotional labor. Acad Manag Rev 21(4):986–1010
19. Stazyk EC, Pandey SK, Wright BE (2011) Understanding affective organizational commitment: the importance of institutional context. Am Rev Public Admin 41(6):603–624

Empirical Study on the Factors Affecting Avatar Usage in Video Conference Platforms

Young-Kun Yang, Gwang-yong Gim, In-Mo Yeo, and Young-Su Kim

Abstract The world has undergone significant changes with the COVID-19 pandemic. While gradually moving away from the influence of COVID-19, remote f working has become normalized, and the work environment is increasingly adapting to remote and online settings. Consequently, the importance of video conference platforms is growing day by day. However, one of the major issues that has emerged with the continued use of video conference plat-forms is Zoom Fatigue. Efforts to address this issue have focused on enhancing privacy protection and improving smooth communication. Nevertheless, research in this area remains limited. Thus, this study aims to conduct empirical study on investigating the intentions behind avatar usage in video conference platforms by utilizing the characteristics of avatars used in the metaverse. Starting with an exploration of avatar attributes, this study identified various organizational perspectives related to avatar usage in smart work and video conference platforms. After conducting a review of relevant theories and prior studies, this study designed a research model and formulated hypotheses accordingly. The research model incorporates the Value-based Acceptance Model (VAM) and the Technology-Organization-Environment (TOE) model, with the independent variables of avatar identification, avatar interaction, avatar anonymity, and technology adoption readiness added as antecedents. Additionally, this incorporated variables from the VAM, such as privacy protection, Perceived Happiness, and sacrifice as benefits, as well as technical effort and innovation resistance as potential variables. To conduct empirical analysis, a survey was conducted targeting employees working in companies that utilize smart work practices. A total of 241 survey responses were used for statistical analysis, after excluding errors and incomplete responses. The

Y.-K. Yang (✉) · G. Gim · I.-M. Yeo · Y.-S. Kim
Graduate Program in IT Policy and Management, Soongsil University, Seoul, Korea
e-mail: yyk0918@naver.com

G. Gim
e-mail: gygim@ssu.ac.kr

I.-M. Yeo
e-mail: visionfor1@naver.com

Y.-S. Kim
e-mail: kys1227@kakao.com

R. Lee et al. (eds.), *AI and Metaverse*, Studies in Computational Intelligence 1160,
https://doi.org/10.1007/978-3-031-61163-6_9

analysis tools employed were SPSS 22 and SmartPLS 4.0. For hypothesis testing, the dedicated structural equation modeling program, SmartPLS 4.0, was used. Lastly, it was confirmed that avatar identification did not significantly influence the mediators. This study has empirical significance as it examines the adoption of avatars, a key element of the metaverse, to address the issues of Zoom fatigue and privacy concerns that have emerged with the increasing user base of video conference platforms after the COVID-19 pandemic. Through empirical analysis, this study validated the key factors for avatar adoption. The study also discusses its limitations and suggests future research directions.

Keywords Video conference platforms · Smart work · Avatar · Privacy protection · Zoom fatigue

1 Introduction

The global COVID-19 pandemic has brought significant changes to society, particularly in the industrial sector (businesses). The implementation of social distancing measures led to the suspension of face-to-face meetings and a sudden demand for remote services. The initial chaotic state of video conference operations has now evolved into the normalization of "everyday video conference." Most users have become accustomed to video conference-related programs, applications, and equipment. Companies now extensively utilize video conference in various areas such as meetings, internal training, seminars, and interviews [1]. Most studies have focused on researching the continued intention to use video conference systems to satisfy users [2].

While the majority of employees and students have quickly adapted to remote work and online classes, there are still many individuals experiencing 'Zoom fatigue,' a phenomenon associated with online video conference. In response to the novel challenges of the COVID-19 era, the scientific community has been conducting various research activities to analyze the causes of Zoom fatigue [3]. Detailed factors contributing to users' video conference fatigue include 'limitations of non-face-to-face communication,' 'difficulty in usage,' 'burden of setting up the surrounding environment,' and 'strain of gaze concentration.' Among these four specific fatigue factors, 'limitations of non-face-to-face communication' and 'burden of setting up the surrounding environment' statistically significantly influenced overall fatigue levels. The factor with the greatest impact was the 'bur-den of setting up the surrounding environment,' which suggests that previously unconsidered factors (such as partial exposure of the user's living space) have a significant influence on overall fatigue experienced by users of video conference services. In particular, university students attending online classes through video conference face the burden of considering various factors and creating a suitable environment [4].

Users of video conference platforms express concerns about surrounding noise and unexpected situations. They feel the need to organize their surrounding environment, which is partially exposed through the screen, in advance, as this effort impacts their overall fatigue levels [5]. Overcoming these limitations, there is a growing need for video conference platforms in smart work environments that utilize metaverses or avatar-based platforms. To alleviate users' discomfort with seeing themselves on video conference systems, avatars have gained popularity as a virtual world solution [5]. Zoom, which captures 50% of the global video conference market, has introduced avatars in the form of virtual animal characters that can be animated based on user movements. Google Meet offers the feature of creating moving 3D avatars, allowing users to combine facial features, skin tones, and hairstyles [6]. Considering that factors such as "exposure to surrounding noise" and the "need for organizing the surrounding environment" emerged as key factors in Zoom fatigue, primarily due to privacy infringements, using avatars can help overcome these challenges to a significant extent.

2 Theoretical Background

2.1 Video Conference Platform

Video conference refers to real-time video sessions between two or more users in different locations. While video conference supports multiple endpoints, terms like "video call" and "video chat" generally refer to one-on-one interactions [7]. With the availability of high-speed internet connections, services like Zoom and Microsoft Teams have made video conference accessible to everyone, without constraints of location or hardware. Academically, video conference is defined as a form of meetings that replaces traditional face-to-face interactions by utilizing the internet to facilitate real-time sharing of video, audio, and data, thus eliminating constraints of time and space for remote education or non-face-to-face meetings [8].

2.2 Perceived Avatar Identification

Perceived avatar identification refers to the strong psychological connection that consumers have with a specific organization or group. It signifies the psychological relatedness or emotional attachment that individuals feel towards a particular entity [9]. Furthermore, perceived avatar identification involves the psychological mechanism where people identify with and emulate the behaviors of a target they aspire to be like [10]. According to the social perception of avatar identification theory, perceived avatar identification represents the strong psychological connection that consumers have with a specific organization or group. The intensity of community perceived avatar identification reflects an individual's psychological attachment to the community [9].

2.3 Perceived Avatar Interaction

It has been found that the social interaction in the metaverse significantly influences trust in both the metaverse platform and organizations associated with the metaverse. Research results suggest that trust formation in the participants of the metaverse is attributed to the social functions provided by the metaverse, and social interaction is identified as one of the key functionalities of the metaverse [11]. The interaction of virtual reality (VR) has been found to have a significant impact on perceived usefulness and Perceived Happiness. This indicates that the interactive nature of VR not only has educational applications but also provides learners with pleasure and enjoyment [12].

2.4 Perceived Avatar Anonymity

In research examining the impact of the degree of anonymity in online com-munities on social conformity and group polarization, it has been found that the degree of social anonymity significantly influences social mimicking conformity, social connectedness conformity, and social comparison conformity, leading to group polarization [13]. A study on the preference for anonymity, network homophily, and individual factors in triggering group polarization in social net-working sites (SNS) found that the preference for anonymity within SNS spaces has a significant impact on in-group conformity behavior, out-group exclusion behavior, and in-group conformity attitudes. It suggests that the stronger the preference for anonymity, the higher the likelihood of group polarization [14].

3 Research Method

This study aims to construct a research model focusing on the intention to use avatars among smart work users who utilize video conference platforms. Considering the organizational perspective inherent in smart work used within companies, this study intends to employ the Technology-Organization-Environment (TOE) model for explanation. Furthermore, to elucidate the intention to use avatars from the perspective of sacrifices and benefits, this study has incorporated the VAM (Value-based Adoption Model) framework (Fig. 1).

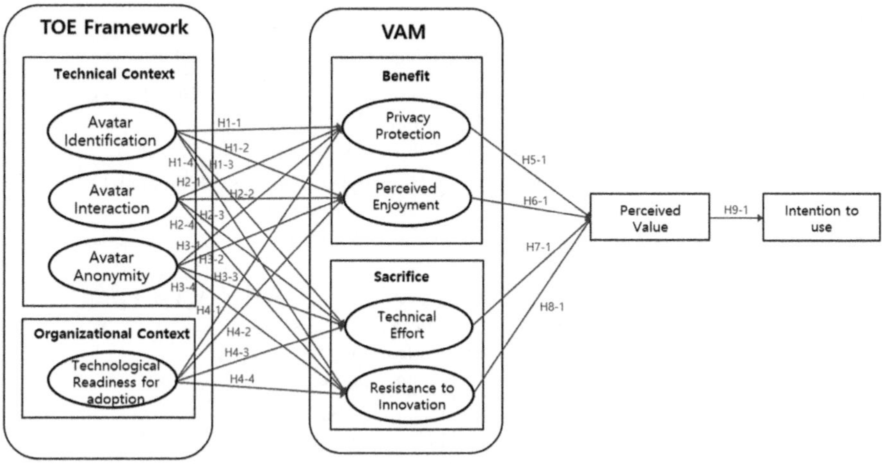

Fig. 1 Research model

3.1 Operational Definition of Each Factor

See Table 1.

4 Empirical Analysis

4.1 Data Collection

To gather the necessary data for the analysis, a questionnaire was developed based on prior research to measure the research model. An online survey was then conducted. The survey targeted office workers who use smart work and have experience using avatars on video conference platforms. The survey was con-ducted from May 26, 2023, to June 1, 2023, for a duration of seven days. A total of 251 survey responses were collected. Among them, 10 responses were excluded due to missing or unreliable answers, resulting in a final dataset of 241 responses for analysis. First, a frequency analysis was performed to examine the demographic characteristics of the participants. Then, reliability and validity analyses were conducted for the measurement variables. Subsequently, advanced statistical analysis was conducted using SmartPLS 4.0 as the analytical tool. The measurement model was evaluated using the PLS algorithm. Following that, boot-strapping was conducted to assess the structural model and test the hypotheses.

Table 1 Operational definition of each factor

Factor	Operational definition of each factor	References
Avatar identification	The act of projecting oneself and feeling psychological connectedness or attachment when using an avatar on the video conference platform	Kim et al. [15]
Avatar interaction	The act of engaging with objects, people, or entities, including oneself, within the given environment when using an avatar on the video conference platform	Shin et al. [16]
Avatar anonymity	The attitude of freely behaving or acting in a manner where identity remains undisclosed or others perceive oneself as unidentifiable among the participants in the video conference platform	Wallace [17]
Technological readiness	The extent to which an organization possesses and can utilize the necessary technological capabilities to adopt innovative technologies in the context of video conference platforms	Oliveira et al. [18]
Privacy protection	The degree of subjective belief that using an avatar on the video conference platform reduces risks to privacy and prevents negative outcomes	Mohamed and Ahmad [19]
Perceived happiness	The level of emotional experience, such as joy or interest, derived from using an avatar on the video conference platform	Davis [20]
Technological effort	The level of non-monetary resources, such as time and effort, expended in using an avatar on the video conference platform	Kim et al. [21]
Innovation resistance	The degree of reluctance or opposition felt when using an avatar on the video conference platform	Reilly and Conover [22]
Perceived value	The overall evaluation of how worthwhile using an avatar on the video conference platform is, encompassing benefits received and sacrifices made	Sirdeshmukh et al. [23]
Usage intention	The extent of willingness or plans to use an avatar on the video conference platform	Bhattacherjee and Sanford [24]

4.2 *Measurement Model for Fit Assessment*

This study applied the Smart PLS structural equation modeling (SEM) to examine the significant path coefficients between variables proposed in the research mod-el and to identify the causal relationships. Furthermore, this study assessed the appropriateness of the model structure and examined the convergent and discriminant validity of each item.

Table 2 Results of the reliability analysis of the model

Factor	Cronbach's alpha	Rho_A	Composite reliability
Technological readiness	0.88	0.882	0.913
Technological effort	0.799	0.867	0.879
Avatar identification	0.863	0.869	0.907
Avatar interaction	0.86	0.865	0.899
Avatar anonymity	0.818	0.826	0.88
Usage intention	0.829	0.831	0.898
Perceived value	0.892	0.893	0.92
Perceived happiness	0.884	0.888	0.915
Privacy protection	0.843	0.854	0.888
Innovation resistance	0.822	0.862	0.88

4.3 Convergent Validity

For all measurement variables, the outer loading values exceeded the threshold of 0.70, and the Average Variance Extracted (AVE) also exceeded the threshold of 0.5. Therefore, it can be concluded that convergent validity has been established.

4.4 Reliability Verification of the Measurement Model

According to the evaluation results, all values of Cronbach's alpha exceeded 0.60, and the rho_A (ρA) values were higher than the threshold of 0.70. The composite reliability (CR) values for all constructs were found to be above 0.60. Therefore, it can be concluded that all variables in the study exhibit satisfactory internal consistency reliability (Table 2).

4.5 Results of Discriminant Validity

Lastly, it is necessary to examine the discriminant validity to assess how well the variables distinguish from each other. To evaluate discriminant validity, the most reliable criterion, HTMT (Heterotrait-Monotrait) ratio, cross-loadings, and HTMT (Heterogeneous and Heterotrait-Monotrait) method from the Fornell-Larcker criterion were employed. The results regarding these criteria are max data is 0.831. So All factor is satisfied un-der threshold 8.5.

4.6 Structural Model Evaluation and Hypothesis Testing

This study analyzed the effects of avatar characteristics (Avatar Identification, Avatar Interaction, Avatar Anonymity) and technological readiness on Privacy Protection, Perceived Happiness, Technological Effort, and Innovation Resistance. Furthermore, this study examined how Privacy Protection, Perceived Happiness, Technological Effort, and Innovation Resistance mediate the relation-ship between the avatar characteristics and Perceived Value. Lastly, this study investigated the influence of Perceived Value on usage intention (Table 3).

The result implies that Avatar Interaction, Avatar Anonymity, and Technological Readiness have an impact on Perceived Happiness and Privacy Protection. Avatar Interaction and Avatar Anonymity are found to influence Innovation Resistance. Additionally, Perceived Happiness, Privacy Protection, and Innovation Resistance are each influencing Perceived Value.

5 Conclusion

This study conducted an empirical investigation on the use of avatars in a smart work environment's video conference platform. Building on previous re-search examining the characteristics of avatar usage, this study approached the topic from an organizational perspective, focusing on the benefits and sacrifices associated with avatar use within a company. To achieve this, the research model was designed based on the TOE frame-work, considering the specific context of a smart work environment and the emerging use of avatars in metaverse platforms. This study aimed to understand the intention to use avatars in video conference platforms within the smart work context, incorporating the unique characteristics of avatars used in metaverse platforms. Additionally, it addressed the growing importance of privacy protection factors in platforms such as Zoom, highlighting their significance in the research.

Data was collected to test the hypotheses of the research model and under-went empirical analysis. Based on the validation results, the following findings were derived.

First, the hypothesis regarding the relationship between Avatar Identification as an independent variable and Privacy Protection, Perceived Happiness, Innovation Resistance, and Technological Effort was rejected. Avatar Identification, which involves the psychological mechanism of aspiring to become like a desired object and immersing oneself in their behaviors, was found to be limited to the extent of individuals shaping their own appearances within the organizational context of the company. It lacked stronger elements of self-projection through avatar's identity and fell short in terms of fulfilling the aspect of identity identification.

Second, the hypothesis regarding the relationship between Avatar Interaction as an independent variable and Technological Effort was rejected. However, it was found to have a significant impact on Privacy Protection, Perceived Happiness, and

Table 3 Results of hypothesis testing

Hypothesis	Original sample (O)	*P* values	Accepted/rejected
Technological readiness → technological effort	−0.181	0.047	Accepted
Technological readiness → perceived happiness	0.143	0.048	Accepted
Technological readiness → privacy protection	0.199	0.001	Accepted
Technological readiness → innovation resistance	−0.091	0.184	Rejected
Avatar identification → technological effort	0.1	0.245	Rejected
Avatar identification → perceived happiness	0.024	0.743	Rejected
Avatar identification → privacy protection	−0.028	0.739	Rejected
Avatar identification → innovation resistance	0.056	0.556	Rejected
Avatar interaction → technological effort	−0.187	0.071	Rejected
Avatar interaction → perceived happiness	0.499	0	Accepted
Avatar interaction → privacy protection	0.189	0.009	Accepted
Avatar interaction → innovation resistance	−0.329	0	Accepted
Avatar anonymity → technological effort	−0.104	0.212	Rejected
Avatar anonymity → perceived happiness	0.107	0.049	Accepted
Avatar anonymity → privacy protection	0.416	0	Accepted
Avatar anonymity → innovation resistance	−0.231	0.002	Accepted
Technological effort → perceived value	−0.042	0.483	Rejected
Perceived happiness → perceived value	0.333	0	Accepted
Privacy protection → perceived value	0.261	0	Accepted
Innovation resistance → perceived value	−0.189	0.002	Accepted
Perceived value → usage intention	0.693	0	Accepted

Innovation Resistance. Users of the video conference platform perceive the act of creating and using avatars to express their thoughts and communicate with others as significant.

Third, the hypothesis regarding the relationship between Avatar Anonymity as an independent variable and Technological Effort was rejected. However, it was found to have a significant impact on Privacy Protection, Perceived Happiness, and Innovation Resistance. This finding is in line with the mentioned factors in the paper, such as the limitations of remote communication and the burden of environmental adjustments, which statistically influence users' overall fatigue [4]. Considering this point, this study contributes to the understanding that using avatars in video conference platforms influences Privacy Protection.

Lastly, it is considered meaningful to investigate the factors influencing the use of avatars in video conference platforms using Virtual Reality (VR), which has recently gained attention. The use of more immersive VR technology can enhance the realism of avatars, providing a significant contribution to the re-search. Furthermore, studying the factors influencing avatar usage from the perspective of technology adoption using Unified Theory of Acceptance and Use of Technology (UTAUT) can validate various factors from a novel technology adoption standpoint.

References

1. Rsupport (2023) Video conferencing market trends. https://files.rsupport.com/kr/remotemeeting/documents/whitepapers/remotemeeting_video_conferencing_trend.pdf. Accessed 14 April 2023
2. Jin S (2021) A study on the effect of perceived quality of video conference software on intention of continuous use: focusing on the mediated effect of user satisfaction—(Ph.D. dissertation). Dankook University, Republic of Korea
3. Weekly Smart (2022). https://www.yna.co.kr/view/AKR20220107136600017. Accessed 14 April 2023
4. Lee E (2022) Video-conference fatigue of college students. J Converg Cult Technol JCCT 8(3):589–594
5. Daily Pop (2021) Video conference, which is essential in the non-face-to-face era, I'm comfortable, but why do I feel more tired?. ttp://www.dailypop.kr/news/articleView.html?idxno=517. Accessed 10 June 2023
6. Ahn H-W, Park J-H, Kim J-Y, Moon J-E, Lee K-M (2023) Video conferencing system using user-specific avatar generation and animation. J Dig Contents Soc 24(5):1061–1069
7. Kim HJ (2022) A study on the factors that influence continuance intention of the video conferencing software: focused on the post acceptance model (Ph.D. dissertation). Kookmin University, Republic of Korea
8. Yujung C (2022) A study on user experience of video conferencing platform, (Master. dissertation). Ewha Womans University, Republic of Korea
9. Ahn K-H, Yoo CJ, Kim S (2023) The study on the structural model of avartar decoration needs, avatar self-identification and site attitude and loyalty. Korea Soc Consum Stud 24(5):1061–1069
10. Hoffner C, Buchanan M (2005) Young adults' wishful identification with television characters: the role of perceived similarity and character attributes. Media Psychol 7(4):325–351
11. Hwang I, Kim J-S, Lee I-H (2022) The effect of innovation and presence factors of Metaverse on the intention to continuous use through platform trust : analysis of the moderating effect of

digital technological competence the Korean academic association of business. Administration 35(6):1055–1079
12. Ranyi B (2021) The effect of VR presence and interaction on education satisfaction in aviation service department education: the mediation effect of perceived usefulness and perceived fun, (Ph.D. dissertation). Sejong University, Republic of Korea
13. Yeon-Woo K (2017) The effects of social conformity on the group polarization according to the anonymity of online community, (Master. dissertation). Hanyang University, Republic of Korea
14. Feng BQ (2014) The effect of anonymity preference, network homogeneity and individual factors on group polarization in SNS, (Master. dissertation). Hanyang University, Republic of Korea
15. Kim C, Lee S-G, Kang M (2012) I became an attractive person in the virtual world: users' identification with virtual communities and avatars. Comput Hum Behav 28(5):1663–1669
16. Shin I, Choi S, Shin H (2011) The influences of elements of the of the interactions in mobile ads on the their acceptance: with the absorption and intrusiveness as the mediating variables. Korea Design Forum
17. Wallace P (1999) The psychology of the internet. Cambridge University Press, Cambridge
18. Oliveira T, Faria M, Thomas MA, Popovič A (2014). Extending the understanding of mobile banking adoption: when UTAUT meets TTF and ITM. Int J Inform Manag 34(5):689–703
19. Mohamed N, Ahmad IH (2012) Information privacy concerns, antecedents and privacy measure use in social networking sites: evidence from Malaysia. Comput Hum Behav 28(6):2366–2375. https://doi.org/10.1016/j.chb.2012.07.008
20. Davis WA (1989) Freedom, enjoyment, and happiness: an essay on moral psychology
21. Kim HW, Chan HC, Gupta S (2007) Value-based adoption of mobile internet: an empirical investigation. Decis Support Syst 43(1):111–126
22. Reilly MD, Conover JN (1983) Meta-analysis: integrating results from consumer research studies. Adv Consum Res 10(1)
23. Sirdeshmukh D, Singh J, Sabol B (2002) Consumer trust, value, and loyalty in relational exchanges. J Market 66(1):15–37
24. Bhattacherjee A, Sanford C (2006) Influence processes for information technology acceptance: an elaboration likelihood model. MIS Quart 805–825

A Study on Technology Trends for AR Glasses Using Topic Modeling

Hyuna Kim, Heeyoung Kim, Gwangyong Gim, and Sungbyeong An

Abstract Augmented Reality (AR) technology is a crucial element in constructing the metaverse and digital twin, as it overlays information on real-world visuals. AR glasses serve as essential tools and devices for delivering such services. This paper aims to review and compare the trends and major technological areas of AR glasses in Korea and the United States by utilizing patent literature and conducting Latent Dirichlet Allocation (LDA) topic modeling. Patent documents already include international standard technology classification codes. However, for comprehensive research and understanding the technological themes from a perspective of convergent studies and technological development, relying solely on these codes might be challenging. Therefore, LDA topic modeling, widely used for semantic topic-based classification of documents, was employed to analyze over 3000 AR glasses-related patents. The analysis of patent documents revealed that Korea focuses on the "application areas of the device," while the United States emphasizes the "optical technology constituting the device".

Keywords AR glasses · LDA topic modeling · Technology trend · Patent

H. Kim (✉) · H. Kim
IT Policy and Management, Soongsil University, Seoul, South Korea
e-mail: hannah_kim@naver.com

H. Kim
e-mail: kkamoong@naver.com

G. Gim
Business Administration, Soongsil University, Seoul, South Korea
e-mail: gygim@ssu.ac.kr

S. An
MIS, Soongsil University, Seoul, South Korea
e-mail: dusung1233@naver.com

R. Lee et al. (eds.), *AI and Metaverse*, Studies in Computational Intelligence 1160,
https://doi.org/10.1007/978-3-031-61163-6_10

1 Introduction

Augmented Reality (AR) technology enriches physical objects and spaces using artificially generated, virtual and synthetic 3D elements created by Computers [1]. Despite the increasing interest in metaverse implementation technology and services, Gartner declared AR as a technology that had moved beyond the hype cycle in 2018 [2]. However, some opinions suggest that AR has become a validated and mature technology [3], potentially forming a new paradigm along with cloud technology [4]. Contrary to Gartner's proclamation, various AR glasses products have emerged since the early 2020s and are being utilized in industrial settings, indicating a need to understand the overall technological trends and trajectories of AR glasses.

Technological trends can be discerned by combining various methods and approaches, such as collecting comprehensive information, analyzing statistical data and Bigdata, seeking advice from experts or specialized institutions, and more. Analyzing patent documents over a specific period is one way to grasp technological trends. Patent literature serves as a rich source of information on emerging technologies [5] and can aid in predicting technological developments in a specific area [6]. Analyzing patent literature for a specific period is also one of the methods to discern technological trends. Patent analysis methods have evolved, and current approaches often involve utilizing natural language processing and text-mining techniques, such as keyword extraction and topic modeling.

This study employs the Latent Dirichlet Allocation (LDA) topic modeling technique, a text mining method, to analyze patent documents and uncover the technological trends of AR glasses. Also by examining the predominant development areas in different countries, we aim to estimate the future direction of AR glasses technology development.

2 Literature Review

2.1 AR Glasses Technology Research

AR and VR headsets are emerging as next-generation interactive displays capable of providing vivid 3D visual experiences. In particular, AR has evolved into the form of glasses by facilitating interaction between users, digital content, and the real world, displaying virtual images while maintaining transparency features. AR glasses involve not only emerging optical technologies [7] but also the integration of complex technologies such as software, content, and various control technologies. Simultaneously, they address challenges related to the weight and comfort of wearable devices. AR glasses are a field where technological development occurs in tandem with addressing the limitations of wearable devices.

Breaking down AR glasses into technological components, the following aspects are considered (Table 1).

Table 1 Technological components of AR glasses

Category	Overview
Display technology	Plays a role in projecting virtual information into the user's field of view, including projection technology, OLED screens, lens technology, etc.
Sensor technology	Includes various sensors for interacting with the user's environment, detecting position, posture, motion, ambient environmental information, etc.
Computing technology	Utilizes computer resources and network connectivity to process and render virtual content in real-time
Control and interaction interface technology	Provides control and input interfaces for users to interact with the virtual environment, including touchpads, voice recognition, gesture control, head movement, and gaze tracking

Table 2 Prior research on technological trends of AR glasses

Title	Author (Year)
Trends of VR/AR wearable display technology	Yang and Kim [8]
Augmented reality device patent trend analysis for highly realistic content experience: focusing on head mounted display augmented reality devices	Jung [9]

Recent prior research on AR glasses has mainly focused on optical-related technologies and usability studies of AR devices. Notable studies on the technological trends of AR glasses include Yang et al., and the research conducted by Jung Dawoon. Yang and Kim [8] investigated the development trends of optical see-through EGD technology, while Jung [9] derived the development trends of head-worn AR device technology through patent analysis (Table 2).

2.2 *Research on Technological Trends in Patents Document Classification Using Topic Modeling*

Research on analyzing patent documents using text mining and artificial intelligence began to emerge in the early 2000s. Nowadays, with the advent of super-sized AI technologies and new language models, studies analyzing and classifying documents, news, search terms, SNS messages, comments on the internet, etc., are proliferating. However, research focusing on patent documents is relatively scarce. This scarcity can be attributed to the absence of openly available and optimized models trained on a massive scale of patent documents. Moreover, dealing with patent documents requires specialized knowledge of their unique characteristics. Currently, topic modeling is the most widely used approach for understanding and analyzing technological trends using documents.

Studies applying topic modeling to analyze technological trends evident in patent documents can be observed in the recent research cases (Table 3). Kim [10] utilized topic modeling to analyze technological trends in the defense science and technology field, Park [11] in the solar energy sector, Park [12] in the factory automation domain, Park and Choi [13] in power semiconductor packaging technology, and Yang et al. [14] in the field of AI.

Latent Dirichlet Allocation (LDA) is one of the algorithms used in topic modeling, a statistical method employed to discover topics within a collection of documents. The functioning of LDA involves initially defining probability distributions for all documents and topics. Topics can be described as sets of words representing themes or main subjects. Subsequently, LDA combines these distributions to estimate the probability of each document having a certain topic. As a result, LDA extracts topics from a given set of documents and analyzes how each document is related to specific topics. This enables the identification and interpretation of themes and patterns in textual data.

In LDA topic modeling, the selection of an appropriate number of topics is crucial for effectively distinguishing technical subjects. Gan and Qi [15] conducted research to calculate the stability of the initial topic model for different topic numbers k, based on the entire dataset, to determine the optimal number of topics. Sbalchiero and Eder [16] presented a clear power-law relationship between the optimal number of topics and the analyzed sample size.

Table 3 Research on technological trends in patents using topic modeling

Title	Author (Year)	Technological trend area
An analysis of technology trends in defense science using data mining techniques	Kim [10]	Defense science and technology
Analysis of technology trends in domestic solar industry through topic modeling	Park [11]	Solar energy
Factory automation trend analysis using latent dirichlet allocation	Park [12]	Factory automation
A study on technology trend of power semiconductor packaging using topic model	Park and Choi [13]	Power semiconductor packaging
A study on analysis of national R&D research trends for artificial intelligence using LDA topic modeling	Yang et al. [14]	Artificial intelligence

3 Research Design

3.1 Data Collection and Status Analysis

For this study, patent documents from South Korea and the United States were collected, focusing on patents published or registered from 2000 to 2022. Approximately 60 Korean and English keywords were utilized for the search. After noise removal from the search results, 752 documents from South Korea and 2,260 documents from the United States were selected for analysis. It's important to note that this search represents a very small fraction, about 10%, of the entire AR technology patent volume, and it is specifically limited to 'Glasses' related technologies.

To understand the overall data landscape, this study examined the number of applications by year, the top 10 applicants, and IPC Sections for both countries (Figs. 1 and 2, Table 4).

Both Korea and the U.S. witnessed a significant increase in patent applications starting from 2010, with rapid growth until 2019. Even considering unpublished patents, the U.S. has experienced a decline in the number of applications since 2019, while Korea is estimated to have started declining after 2021 (Fig. 1).

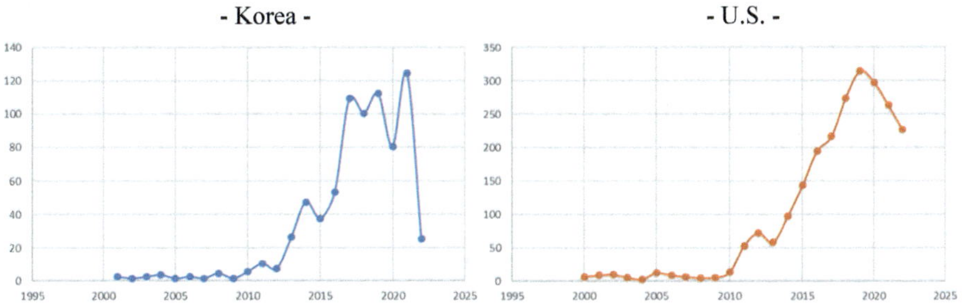

Fig. 1 Yearly trends in patent applications of AR glasses technology

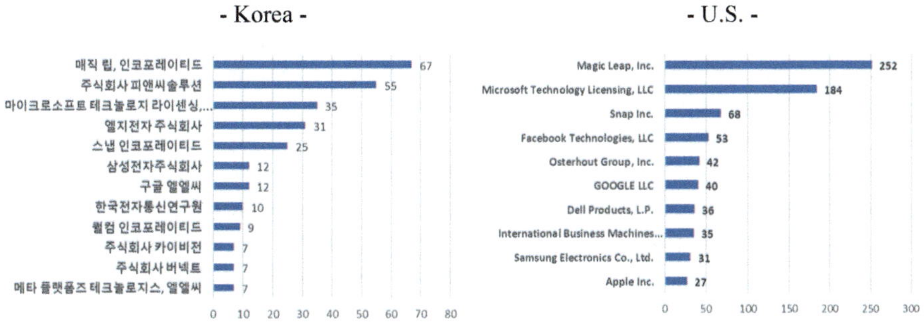

Fig. 2 Top 10 patent applicants in AR glasses technology

Table 4 Number of cases by IPC section of AR glasses technology

IPC section	Section title	Korea	U.S
A	Human necessities	95 (12.6%)	176 (7.8%)
B	Performing operations; transporting	13 (1.7%)	16 (0.7%)
F	Mechanical engineering; lighting; heating; weapons; blasting	2 (0.3%)	16 (0.7%)
G	Physics	597 (79.4%)	1,837 (81.3%)
H	Electricity	45 (6.0%)	215 (9.5%)

In Korea, all six of the top 10 applicants are global multinational companies, indicating a high reliance on foreign technology. In the U.S., Samsung Electronics is the only Korean company among the top applicants, and the number of patent applications from Samsung Electronics is about three times greater than those from Korea. Both Korea and the U.S. have Magic Leap as the top applicant (Fig. 2).

The IPC Section, representing the technology classification, is overwhelmingly dominated by the G section in both Korea and the U.S. The G section encompasses a broad spectrum of technology, dealing with the engineering aspects of hardware devices, software algorithms, and component technologies for systems. Notably, Korea shows a significant proportion in the A section, indicating the utilization of AR glasses, while in the U.S., the H section, representing electrical/electronic/telecommunication technologies related to AR glasses devices, takes the second position. This difference is noteworthy (Table 4).

3.2 Analysis Methods and Procedures

First of all, when preparing the datasets for Korea and the United States, the text was standardized to English to eliminate language differences. Although there are studies on cross-language LDA, such as Cross-Language LDA [17], for comparing across multiple languages, this experiment used the basic model and library. Standardizing to English helps avoid potential variations in modeling results based on language structure and grammatical characteristics.

Next, selecting the model and setting parameters, the most challenging part is determining the optimal number of topics. To address this, perplexity and coherence were calculated during the modeling process. Perplexity indicates how well the model describes and predicts the given data, with lower values suggesting better model performance. Coherence evaluates how semantically consistent the topics are, and a higher coherence score indicates that the model has extracted meaningful topics effectively. However, it's not straightforward to determine the optimal number of

Table 5 Research procedures

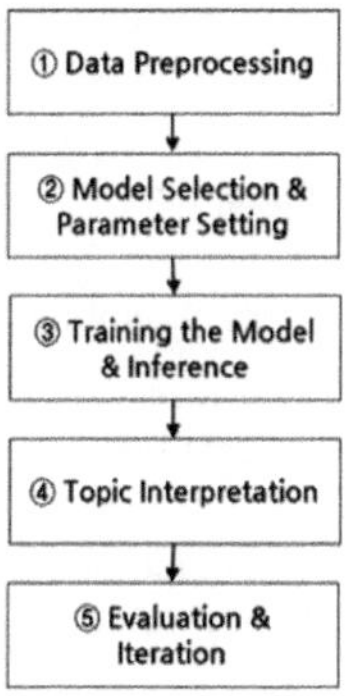

① Data preprocessing for four columns (summary, application year, application number, IPC section) to be used in the analysis
② Selecting a model and setting parameters
③ Training the model and inferring topics (internally)
④ Interpreting topics using derived keywords
⑤ Evaluating results and iterating the modeling process

topics based solely on these metrics. Because a low level of Perplexity does not necessarily guarantee a high level of Coherence (Table 5).

In this study, in addition to Perplexity and Coherence, Intertopic Distance Map was considered, and the optimal number of topics was set to 7.

3.3 Analysis Results

3.3.1 Overview of Topics in Patents Documents by Country

To understand the specific content of the topics, the top 50 keywords were extracted to interpret the technology of each topic. The results are as follows (Tables 6 and 7, Fig. 3):

Using the probabilities of each topic contained in each document, we can calculate the sum of each topic to determine the proportion of each topic in the dataset. Through this topic proportion, we can understand the main research and development areas for each country.

For Korea, topics 3 and 6 have the highest proportions. Topic 3 mainly addresses methods of interacting with virtual objects using devices, while Topic 6 focuses on the application areas of devices (Table 6). Examining the time-series trends for each topic, as shown in the graph above, topics 2, 4, 5, 6 are decreasing in the latter part of the analysis period, while topics 0, 1, and 3 continue to show fluctuations (Fig. 3).

In the case of the United States, topics 1 and 5 have the highest proportions. Topic 1 focuses on interaction with virtual worlds using devices, and Topic 5 emphasizes optical systems for constructing devices (Table 7). Looking at the time-series trends for each topic, as shown in the graph above, topics 2, 4, and 5 are decreasing in the latter part of the analysis period, while topics 0, 1, 3, and 6 continue to show fluctuations (Fig. 3).

Table 6 Overview of 7 Topics in Korea Patent Documents

Topic No.	Topic analysis using keywords
0	Technology related to systems for actual worker training using smart head-mounted devices
1	Technology related to new wearable head-mounted devices integrating optical sensors and gesture control functions
2	Technology related to optical head-mounted devices, including dual-lens assemblies for medical professionals
3	Technology related to providing mixed reality experiences by delivering virtual content based on user's environment and location
4	Technology related to glasses with depth (camera) sensors and advanced motion recognition capabilities, serving as an interface for user hand gestures and voice input
5	Technology related to devices and methods allowing viewing images using optical systems
6	Technology related to systems providing immersive user experiences by mixing real and virtual elements in various domains, including multi-dimensional mixed reality systems

Table 7 Overview of 7 topics in U.S. patent documents

Topic No.	Topic analysis using keywords
0	Encompasses various technologies for using user head-mounted devices
1	Technology related to integrating virtual objects related to the user's physical space using HMD, dynamically interacting with the real world
2	Technology related to stereoscopic image processing for health applications using wearable glasses
3	Technology related to high-quality interfaces for users to interact with virtual environments using HMD
4	Technology related to 3D surgical imaging systems for precise medical procedures
5	Technology related to optical waveguide systems for enhanced imaging and projection
6	Technology related to devices configured to process and display information to users based on their gaze

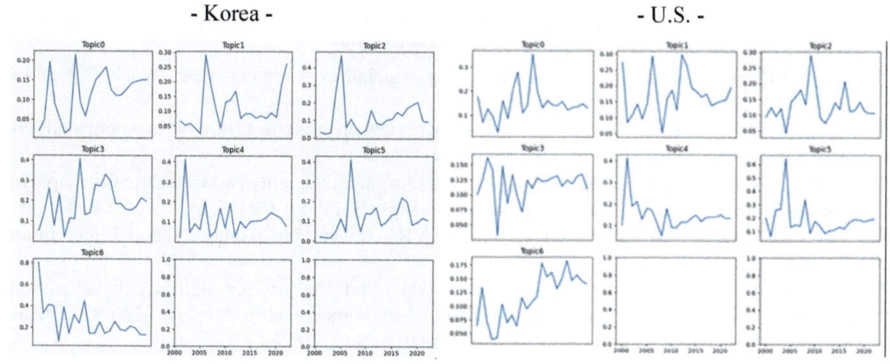

Fig. 3 Time series trends of patent document by topic

4 Conclusion

This study analyzes the technological trends of AR Glasses using Patent Documents and LDA topic modeling. By investigating the evolution and trends of technology over the past 22 years from 2000 to 2022, the study aims to understand the flow of AR Glasses technology and identify similarities and differences between South Korea and the United States.

According to the research findings, both South Korea and the United States focus on the technology related to the combination of the real world and virtual content and interaction in AR Glasses. At the national level, South Korea emphasizes topics related to the application field of the device, while the United States focuses on topics associated with the construction of optical systems. The study also suggests a development trend towards using user gestures or gaze as input interfaces of AR Glasses.

From a research methodology perspective, this study highlights that text mining techniques can be used to easily understand the topics of large numbers of patent documents. Additionally, it emphasizes the importance of utilizing Perplexity and Coherence with Intertopic Distance Map to determine the optimal number of topics.

In future research, the inclusion of representative Claims in addition to the summary column could lead to a more detailed extraction of technological topics in narrower areas.

References

1. Altinpulluk H (2017) Current trends in augmented reality and forecasts about the future. In: ICERI2017 proceedings. IATED, pp 3649–3655
2. https://www.gartner.com/en/newsroom/press-releases/2018-08-20-gartner-identifies-five-emerging-technology-trends-that-will-blur-the-lines-between-human-and-machine
3. https://arpost.co/2020/09/25/augmented-reality-gartners-hype-cycle/

4. Muñoz-Saavedra L, Miró-Amarante L, Domínguez-Morales M (2020) Augmented and virtual reality evolution and future tendency. Appl Sci 10(1):322
5. Ernst H (2003) Patent information for strategic technology management. World Patent Inf 25(3):233–242
6. Abbas A, Zhang L, Khan SU (2014) A literature review on the state-of-the-art in patent analysis. World Patent Inf 37:3–13
7. Xiong J, Hsiang EL, He Z, Zhan T, Wu ST (2021) Augmented reality and virtual reality displays: emerging technologies and future perspectives. Light Sci Appl 10(1):216
8. Yang O, Kim K (2016) Technological trends in VR/AR wearable display. J Electr Telecommun Trend Anal 31(4):1–10
9. Jung D (2021) Augmented reality device patent trend analysis for highly realistic content experience: focusing on head mounted display augmented reality devices. Master's Thesis, Soongsil University Graduate School of Information Science, Seoul
10. Kim (2023) An analysis of technology trends in defense science using data mining techniques. Master's Thesis, Korea University Graduate School of Computer Information and Communication, Seoul
11. Park (2023) Analysis of technology trends in domestic solar industry through topic modeling. Master's Thesis, Korea University of Technology and Education Graduate School, Cheonan
12. Park (2022) Factory automation trend analysis using latent dirichlet allocation. Master's Thesis, Sunmoon University Graduate School, Asan
13. Park K, Choi G (2020) A study on technology trend of power semiconductor packaging using topic model. J Microelectron Packag Soc 27(2):53–58
14. Yang et al (2021) A study on analysis of national R&D research trends for artificial intelligence using LDA topic modeling. J Internet Comput Service 22(5):47–55
15. Gan J, Qi Y (2021) Selection of the optimal number of topics for LDA topic model—taking patent policy analysis as an example. Entropy 23(10):1301
16. Sbalchiero S, Eder M (2020) Topic modeling, long texts and the best number of topics. Some problems and solutions. Qual Quant 54:1095–1108
17. Chang CH, Hwang SY (2021) A word embedding-based approach to cross-lingual topic modeling. Knowl Inf Syst 63(6):1529–1555

A Study on the Derivation of Digital Citizenship Components and the Development of Educational Curriculum

Jin-Hee Son and You-Kyung Lee

Abstract Digital citizenship consists of a knowledge area based on a basic understanding of the digital society, a functional area including digital utilization and thinking ability, and a value and attitude area for correct participation in the digital society. This study aims to derive the concept of digital citizenship and the components of digital citizenship education through literature research and in-depth interviews with experts. The draft digital citizenship curriculum first selected 21 topics under three categories of knowledge, function, value, and attitude through literature research. The researchers conducted the first expert interview on the appropriateness and validity of 21 topics and content elements, and teaching and learning programs. Experts presented various opinions on the revision and integration of digital citizenship-related topics and content elements for each topic as researchers related to digital citizenship and instructors at civic universities. As a result, 10 topics were selected from 21 topics, and a digital citizenship curriculum revision was developed based on these 10 topics. Afterwards, through the second expert interview, opinions on the digital citizenship curriculum module, content element modification, teaching and learning methods, and application of the curriculum were collected and reflected in the final draft of the digital citizenship curriculum.

Keywords Digital society · Digital citizenship · Digital citizenship components · Digital citizenship educational curriculum

---

J.-H. Son
Department of Youth Coaching and Counseling, Korea Soongsil Cyber University, Seoul, South Korea
e-mail: ssony@mail.kcu.ac

Y.-K. Lee (✉)
Department of Christian Counseling and Welfare, Korea Soongsil Cyber University, Seoul, South Korea
e-mail: hontas1@mail.kcu.ac

R. Lee et al. (eds.), *AI and Metaverse*, Studies in Computational Intelligence 1160,
https://doi.org/10.1007/978-3-031-61163-6_11

1 Introduction

The development of digital technology has brought great changes to civil society. The formation of a digital society consists of the development of a working technology system in the first stage, the formation of a legal and institutional foundation in the second stage, and the development of information services to improve citizens' knowledge and attitude skills in a digital society in the third stage [1]. In light of this stage of digital society formation, our reality is focused on creating an external environment, such as the formation of policy stages, such as the provision of legal and institutional devices necessary for the digital society, and the education of citizens, the main body of the digital society, is relatively neglected. Citizens in a digital society are quickly provided with information and knowledge on political and social issues, express interest in a wide range of areas according to personalized values, and participate in the digital society through various means. Citizens in the digital society reflect the self-image of the real society, whether conscious or unconscious. In other words, citizens in a digital society are a reflection of a realistic self, and digital citizens cannot exist without a realistic self. Therefore, in a digital society, citizens need a sense of responsibility, ethics, and an attitude for a smooth relationship with others based on digital literacy. Ribble emphasized that digital citizens must act appropriately and responsibly in using technology, and that it is necessary to cultivate citizenship required in the digital society to live in harmony with other community members as citizens of the digital society [2]. In response to these demands of the times, education is needed to acquire the qualities required of citizens in a digital society.

Despite this importance, education on digital citizenship is focused on regular school education. According to the case analysis of digital citizenship education in Korea, education that supports teenagers to grow into healthy citizens and use information worthily in a digital environment, communication and relationships, personal information and reputation, information selection and management, education that deals with social participation, online bullying, and digital etiquette are dominated by education for elementary, middle, and high school students, including development of learning topics that can be used in connection with achievement standards by elementary, secondary school level, and subject [2]. Recently, the government and various related organizations are attempting digital citizenship education, but improvement is required due to the lack of a systematic educational approach that encompasses the knowledge, skills, values, and attitudes required of digital citizens.

Therefore, it is necessary to identify the concept of digital citizenship in line with the changing trend of society, derive digital citizenship education components, and present a systematic education plan. To this end, this study aims to develop a specific curriculum on the contents of digital citizenship education for citizenship education after the regular curriculum. In addition, we would like to present a specific operation plan for the effective digital citizenship curriculum for fostering digital citizenship.

2 Theoretical Background

2.1 *The Concept of Digital Citizenship*

Digital citizens are those who actively and responsibly participate in online and offline communities in regions, countries, and the world through extensive capacity development [3]. From this point of view, digital citizenship can be seen as a quality required of citizens operating in a digital society. The concept of digital citizenship contains various characteristics of existing traditional and critical approaches such as social responsibility, literacy ability, and active participation. Therefore, citizenship in a digital society needs to be understood along with the existing concept of citizenship of the traditional approach [4], which is not a fixed concept, but expands or changes according to social changes.

Prior research on digital citizenship sees digital citizenship as the ability necessary in the digital environment, the norms of conduct to be observed in the digital environment, and the ability to freely participate in society in the digital environment [5]. The discussion of the concept of digital citizenship is expanding from basic digital utilization abilities to citizens' abilities necessary to communicate and participate in the digital environment. Kim et al. [6] saw that cognitive information processing ability, democratic tendencies, decisions, and behavioral abilities are a kind of ability and quality that combine knowledge, skills, and attitudes in relation to digital citizenship, and that there is a responsibility to handle the seemingly infinite freedom on its own. Park [7] believes that digital citizenship is differentiated in that it is based on communication using ICT, and defined digital citizenship as free expression of opinions online, respect for other people's opinions and thoughts, support or opposition to public issues, manners to keep in expressing one's opinions or interests, and public consciousness. Kim et al. [8] believe that the discussion on digital citizenship presented so far varies depending on whether it is viewed simply as an online society or as a whole in which digital technology is used, but in general, it includes all the skills, attitudes, and behavioral factors that people who use digital technology should have. Kim [9] saw digital citizenship as a dynamic and flexible concept linked to an individual's daily online and offline activities, not static, stable and fixed, and saw active participation in various interests and problems across digital and real society as the biggest feature. It can be seen that digital citizenship includes not only the use of digital devices, but also the knowledge, skills, values, and attitudes necessary to live as citizens in a digital society. The concept and components of digital citizenship are not limited to one area, but are characterized by active participation, rational communication, and interaction based on digital information literacy that can be used in a digital environment-based society [10]. Putting these discussions together, digital citizenship can be seen as the ability to actively and responsibly participate through digital devices based on an understanding of the digital society, and to reflect on the impact of one's activities in the digital society.

2.2 *Digital Citizenship Components*

In this study, the components of digital citizenship were classified into knowledge, function, values, and attitudes through the analysis of domestic and foreign prior studies related to digital citizenship. The specific areas and main content elements are as follows.

2.2.1 Knowledge

Among the components of digital citizenship, knowledge focuses on a basic understanding of the digital society. It recognizes changes in digital democracy and digital society, and includes an understanding of rights and responsibilities within a digital society (Table 1).

Table 1 Digital citizenship components: knowledge

Area	The main contents
Digial democracy	• Solving political, economic, social, and cultural problems arising within the existing society and demanding changes in the social system • Express and participate in traditional power and inequality organizations in digital society
	References: [4, 11, 12]
Digital social understanding	• Awareness of political, economic, cultural, social, and educational issues affecting the digital society • Understanding the Fourth Industrial Revolution (AI and IoT, Robotics, etc.) • Sensitivity to community and global issues, ethical understanding of information covering community and global issues
	References: [4, 8, 11, 13]
Digital social rights	• Recognize rights in a digital society and understand that everyone has equal access to digital rights • Ensure digital rights and access opportunities by ensuring equal opportunities and consideration for the digital underprivileged due to poverty, physical disabilities, physical situations, etc
	References: [14–17]
Digital social responsibility	• Safe, ethical and responsible use (electronic responsibility for actions) • Respect other people's digital rights, including copyright, plagiarism, hacking, privacy, password security, and data backup
	References: [2, 4, 6, 14–16]

2.2.2 Function

Among the components of digital citizenship, functions range from rudimentary device manipulation to content development using it, and include not only technical skills but also the thinking skills necessary to use them (Table 2).

2.2.3 Values and Attitudes

Among the components of digital citizenship, values and attitudes are to understand one's identity as a digital citizen, to know the cultural norms of digital society and to respect basic values. It is also possible to participate in various areas of the digital society with a balanced attitude and reflect on their activities in the digital society (Table 3).

3 Research Method

Through literature research and several consultations with researchers, digital citizenship-related competencies were selected into 21 areas under three categories: knowledge, function, value, and attitude, and key content elements were derived accordingly. Five experts were selected for digital citizenship content elements and curriculum development advice. Judgment data on the importance of digital citizenship-related areas and key content elements were delivered to experts before the expert interview (Table 4).

After the first expert interview to design core civic competencies in the digital citizenship curriculum, researchers organized a curriculum focusing on core civic competencies. In order to examine the validity and direction of this, the second expert interview was conducted with related experts such as civic university instructors, lifelong education experts, and social studies education professors. Through this, the purpose of the interview, the schedule and target, and the main contents of the interview are as follows, as we tried to verify and modify the validity of the curriculum (Table 5).

4 Reviewing Digital Citizenship Components

4.1 Components of Digital Citizenship's 'Knowledge' Area

Important competencies in the area of knowledge presented in expert interviews conducted in this study included 'digital social rights', 'digital social understanding', 'digital social responsibility', and 'digital democracy'. It was suggested that in digital

Table 2 Digital citizenship components: function

Area	The main contents
Ability to utilize digital devices	• The ability to use technology and tools used to exchange, disseminate, share, and receive information (knowledge) • Ability to find and use digital technology tools effectively and efficiently • Recognizing that opportunities to use digital technology as a participant in digital society may not be fair to everyone
	References: [2, 4, 12, 14, 16, 18]
Digital information access capabilities	• The ability to manipulate digital to obtain the information you want on digital (electronic access)
	References: [8, 19, 20]
Digital information evaluation capabilities	• Ability to critically evaluate and apply digital information • Ability to evaluate the reliability of a variety of information and sources • Ability to analyze information quality, truthfulness, reliability, perspective, etc. through critical thinking • The ability to consider the potential influence or importance of information
	References: [8, 21–23]
Digital contextual thinking skills	• Based on the convergence and reality of the digital society, we think flexibly according to the situation the ability to reasonably solve problems • The ability to interpret or construct dynamic real-world situations and processes in a digital society
	References: [8, 22, 24]
Digital safety management capabilities	• Knowing and observing electronic precautions for security • Ability to distinguish between secure websites • Inadequate access and access, the ability to distinguish information
	References: [2, 14, 16, 19, 20]
Digital privacy capabilities	• Maintain password security, back up data, and set appropriate privacy on browsers and social media platforms • Protect your data using anti-virus software or firewalls for privacy
	References: [2, 14, 16, 20, 25]

(continued)

Table 2 (continued)

Area	The main contents
Digital trace management capabilities	• The ability to recognize that actions in a digital society leave a mark, to follow etiquette and laws, and to manage them • The ability to understand the permanence of the digital world and to proactively manage personal information
	Reference: [22]
Digital communication skills	• The ability to exchange digital information (knowledge) electronically • The ability to know and use different digital methods of communication • The ability to express and communicate your thoughts and opinions digitally in multiple ways
	References: [8, 11, 12, 15, 16, 18, 19, 26]
Digital creative ability	• The ability to make your message digitally understandable to others • The ability to create new knowledge digitally and to reconstruct and reinterpret knowledge derived from various fields
	References: [22, 24, 27]
Digital collaboration capabilities (digital networking capabilities)	• The ability to gather digital information and exchange views with others for a common purpose • Networking (relationship formation) ability to develop relationships with others through information sharing and collaboration using various digital media
	References: [4, 6, 16, 27]

social rights, the guarantee of rights to the underprivileged, that everyone should get equal opportunities in a digital environment, and education on basic rights such as copyright and privacy protection are necessary. In understanding digital society, basic situations and citizenship education on computer use, understanding basic technology, understanding data management, and sensitivity to global understanding or global issues were considered important. There was an opinion that it was necessary to recognize the importance of observing the etiquette of the digital society and respecting the rights of others as a digital society responsibility. It also emphasized that as much as one's personal information protection is important, the personal information of others should be respected and considered important. Regarding digital democracy, it mentioned the need to relate to civic participation such as national petitions and the fact that politicians' posts or posts in digital political, economic, social, and international activities have equal influence when comparing posts (posts) posted by ordinary citizens. In addition, it emphasized the need to understand technological changes such as coins, blockchain, and cryptocurrency, problems arising from

Table 3 Digital citizenship components: values and attitudes

Area	The main contents
Digital social identity	• Trying to establish the identity of the digital world • Learn how to harmonize digital and analog life and strive • Efforts to apply the right attitude in a digital society, with the belief that one can reasonably change one's views by evaluating one's own prejudices
	References: [11, 16, 28]
Understanding digital socio-cultural norms	• Embrace and understand the diverse cultural norms of the community through various digital media • Understanding cultural differences in the community of digital society • Interacts with diverse communities in the digital society, respecting and acknowledging communities of different values
	References: [22, 27]
Respect the basic values of digital society	• Respect basic values in a digital society • Empathize and communicate with others using digital for public purposes
	Reference: [11, 16, 28]
Health in a digital society	• Physical and mental health in a digital society • Balancing digital usage • Balance the digital society and the real world • Recognizing digital addiction
	References: [2, 14–16, 21, 22]
Civic engagement in digital society	• Participation in actively sharing one's opinion or interest in writing in a digital society • Appropriate participation as a digital citizen at the local, national, and international level to solve problems • Participation in various areas of digital society, including political, social, economic, and cultural participation
	References: [2, 4, 11, 12, 23, 26, 28]
Digital social reflection	• Reflectively reflect on one's own identity, communication, life experience, and behavior in the digital society by considering the responsibilities and ethical principles of the digital society
	References: [18, 23]

Table 4 Overview of conducting the first expert interview

Categorized	Content
Goal	• Modification of the domain and content elements of the digital citizenship curriculum developed by literature research • Modification and supplementation of digital citizenship curriculum • Field appropriateness review of digital citizenship curriculum
Schedule	• Date and time: May 16, 2022 (Monday)–May 19, 2022 (Thursday) • Method: Utilize ZOOM
Target	• Five people, including researchers on the concept of digital citizenship and instructors on digital-related experience such as civic universities
Content	• Opinion on the competence and domain of digital citizenship curriculum • Opinion on the content elements of the digital citizenship curriculum • Opinion on how to operate education in digital citizenship education programs

Table 5 Overview of conducting the Second expert interview

Categorized	Content
Goal	• Verification of validity and field applicability of the first developed digital citizenship curriculum • Complementary to digital citizenship curriculum revision • Improving the Field Appropriateness of Digital Citizenship Curriculum
Schedule	• Date and time: June 16, 2022 (Thursday)–June 19, 2022 (Sunday) • Method: written (email document) answer method
Target	• Four people, including civic university instructors and lifelong education experts
Content	• Opinion on digital citizenship curriculum area • Opinion on the topic and method of education in the digital citizenship curriculum • Opinions on the digital citizenship education plan and module

them, biased news through digital algorithms, and biased information. Considering the word citizenship, it was suggested that it is more important to understand the social context or structure and know the rights of members within it than functional parts. Through expert interviews and consultations with researchers, digital social understanding, digital social rights, and digital social responsibility were selected as content elements related to "knowledge" in the digital citizenship curriculum. Through reflection of experts' opinions and further discussions, the researchers moved digital social responsibility to the value and attitude areas among the areas currently in the knowledge area. The main content elements were finally selected and presented after moving, modifying, deleting, and integrating between areas.

4.2 Components of Digital Citizenship's 'Function' Area

Important competencies in the functional area were "digital device utilization ability," "digital collaboration ability," "digital communication ability," "digital information evaluation ability," "digital contextual thinking ability," "digital trace management ability," "digital personal information protection ability," and "digital information access ability." What should be considered important as a content element is that not only the ability to use the device, but also how personal information is not exposed is necessary. And it was suggested that digital collaboration can be combined with digital communication, and that the older generation and the younger generation need to teach the younger generation's digital culture to close this gap because the difference in digital culture is large. In addition, the importance of the ability to evaluate the spread of fake information or false news obtained from non-professionals during information search has grown, and the practice of professional and reliable information search ability is proposed because of the large difference in individual information search and utilization ability. Recently, he emphasized the contextual thinking ability to connect parts that are necessary or important by examining the digital environment and circumstances, seeing that contextual thinking is important to know the source of various information and how this information is shared in relation to digital citizenship. Digital citizenship 'function' related content elements were digital information access and utilization (two areas integrated), digital information evaluation, digital safety and personal information protection (two areas integrated), and digital communication and creation ability (two areas integrated areas).

4.3 Components of Digital Citizenship's 'Values and Attitudes' Area

Important competencies in the area of "values and attitudes" were "understanding digital social and cultural norms," "respecting basic values of digital society," and "reflection of digital society." Important content elements to be dealt with in the area of values and attitudes were the way citizens communicate in a digital society, reflecting on what they express and act online, understanding cultural norms, and identity. It was considered that different cultures exist in different communities in the digital society, the way they respect them is important, and the attitude to listen to the opinions of people who speak differently from themselves is important. It also emphasized the need for efforts and attitudes for people to actively participate in solving problems on various media platforms in the digital society. The final selection of values and attitudes related to the digital citizenship curriculum was "understanding the cultural norms of digital society," "respect and responsibility for the values of digital society," "digital health," and "participation and reflection of digital society." Another important point in the area of values and attitudes related to digital citizenship is that it is necessary to recognize that the development of digital technology is not always negative. The development of digital technology has two sides, so it is important for people to pay attention to the negative aspects relate.

5 Composition of Digital Citizenship Curriculum

There were a total of 10 topics derived from digital citizenship education programs, divided into two in the knowledge area, four in the functional area, and four in the value and attitude area. Two knowledge areas (change of digital society, rights of digital society), four functional areas (access and utilization of digital information, digital information evaluation, digital safety and personal information protection, digital communication and creative ability), and four values and attitudes (understanding cultural norms, respect and responsibility of digital society, digital health, participation and reflection of digital society) were finally selected. The content elements of each of the 10 topics are briefly summarized as follows (Table 6).

6 Conclusions

Our society is closely related to the digital society, and the citizens of the digital society reflect the citizens of the real society. Therefore, it is necessary to cultivate the qualities to be active and responsible in a digital society. In a digital society, citizens should be able to participate actively and responsibly in a digital society based on their understanding of it. Although education is required to foster such digital citizenship, digital-related education is currently mainly conducted within school education, and digital-related education for ordinary citizens has a limitation in that it focuses only on the use of digital devices. Therefore, education is required to cultivate both the knowledge aspect of the digital society, the functional aspect necessary for participating in the digital society, and the value and attitude aspect of participating in the digital society with a responsible attitude.

As a result of examining digital citizenship education in Korea, attempts have recently been made to actively practice digital citizenship education, but it is necessary to reflect the latest issues and balance knowledge, functions, values, and attitudes. In addition, it is required to construct a curriculum that allows active interaction rather than unilateral education. Looking at overseas digital citizenship education, it was found that it was necessary to cultivate functions through digital citizenship education, develop a digital citizenship curriculum suitable for our social reality, and develop a curriculum for ordinary citizens. In this study, in order to solve the problems presented above, a curriculum was intended to cover the knowledge, function, values, and attitudes of digital citizenship and to actively participate in learners and enhance citizens' digital capabilities. Digital democracy, digital social understanding, digital social rights, and digital social responsibility were derived as knowledge areas, and digital device utilization, digital information access, digital information evaluation,

Table 6 Digital citizenship education content elements

Topic	The main contents
The transformation of the digital society	• Understanding the fourth industrial revolution • Awareness of political, economic, cultural, social, and educational issues affecting the digital society • Problems and reflections on digital society changes
Digital social rights	• Recognition of a number of equal rights related to the digital society • Understanding the digital underprivileged and giving them equal opportunities for consideration
Ability to access and utilize digital information	• Ability to efficiently find and use digital technology tools • The ability to manipulate digital to get the information you want on digital • Recognizing that opportunities to use digital technology as a participant in digital society may not be fair to everyone
Digital information evaluation capabilities	• The ability to critically evaluate and apply digital information • Analyzing information quality, truthfulness, reliability of sources, perspective, etc. through critical thinking critically • Judging digital information through critical thinking to identify unauthenticated materials and not sharing them
Digital safety and protection capabilities	• Knowing and observing electronic precautions for security • Secure websites, improper access and access, and the ability to distinguish information Maintaining password security, backing up data, setting up appropriate privacy on browsers and social media platforms • The ability to recognize that actions in a digital society leave traces, to act in accordance with etiquette and related laws, and to manage them
Digital networking and creativity	• The ability to collect digital information and interact with others while participating in activities • Networking (relationship-building) ability to leverage digital media to build and develop interpersonal relationships • Ability to understand and use different digital communication methods • The ability to create, reconstruct and reinterpret original content

(continued)

Table 6 (continued)

Topic	The main contents
Understanding the cultural norms of the digital society	• Embrace and understand the diverse cultural norms of the community through various digital media • Understanding cultural differences in the community of digital society • Interacts with diverse communities in the digital society, respecting and acknowledging communities of different values
Respect and responsibility for the value of digital society	• Exploring ways to understand basic values and respect values in a digital society • Exploring digital utilization for public purpose • Empathize and communicate with others' values and behaviors in a digital society • Safe, ethical and responsible use of digital • Respect for the rights of others
Digital health	• Physical and mental health in a digital society • Balancing understanding media addiction and digital usage • Understanding a harmonious life between the digital society and the real world
Digital society engagement and reflection	• Participation in actively sharing one's opinion or interest in writing in a digital society • Freely form rational judgments and opinions on public issues in the digital society • Proactively participate in solving problems culturally, democratically, and politically within a digital society • Reflectively reflect on one's own identity, communication, life experience, and behavior in the digital society, thinking about the responsibilities and ethical principles of the digital society

digital contextual thinking, digital safety management, digital privacy, digital trace management, digital communication, digital creation, and digital collaboration were derived as functional areas. In the area of value, the identity of digital society, understanding digital social and cultural norms, respect for basic values of digital society, health in digital society, participation of digital citizens, and reflection of digital citizens were derived through research.

Based on the results of this study, the suggestions are as follows. First, digital citizenship is not a fixed concept, but a concept that continues to expand and develop. Therefore, the digital citizenship curriculum developed through this study needs to be updated to meet the development of technology and the needs of the times. To this end, it is necessary to continuously analyze issues and problems in the digital society and develop the program by reflecting them. In addition to activating the monitoring of educational programs, the revision of the curriculum and the development of educational programs in line with the pace of expansion of digital citizenship should be carried out in the mid to long term.

References

1. Yun SY, Lee WT (2009) A policy recommendations for improving digital citizenship. J Korean Assoc Reg Inform Soc 12(1):59–84
2. Ribble M (2011) The nine elements of digital citizenship, digital citizenship in schools, 2nd edn., pp 16–42
3. Council of Europe (2022) Digital citizenship education handbook
4. Choi MS, Park HJ (2015) A validation study on the Korean digital citizenship scale through exploratory and confirmatory factor analysis. Theory Res Citizen Educ 47(4):273–297
5. Kim AM, Lee JY, Ju JJ, Lee YJ, Yang SE (2019) A study on the concept of digital citizenship and educational methods. Gyeonggi Institute of Education
6. Kim BS, Kim HC, Park SA, Im SS (2017) Exploring the digital citizenship of intelligent information society in the era of the fourth industrial revolution. KERIS
7. Park SH (2020) Using of digital textbook for the cultivation of digital citizenship. J Digit Converg 18(2):111–119
8. Kim BY, Kwon HY, Kim MR (2021) A study on the improvement of educational system to strengthen digital citizenship in the age of artificial intelligence. J Korean Assoc Comp Educ 24(3):67–88
9. Kim YH (2021) Social studies educational implications of digital citizenship. J Learner-Center Curricul Instruct 21(19):127–144
10. Lee J, You SK, Lee YO (2020) Needs and perceptions of middle and high school students for digital citizenship education. Korean J Teach Educ 36(4):123–144
11. Cho IS (2009) A study on digital citizenship in the normative perspectives. J Korean Assoc Reg Info Soc 12(1):11–29
12. Choi M (2016) A concept analysis of digital citizenship for democratic citizenship education in the internet age. Theory Res Soc Educ
13. Community Media Foundation (2021) Best practices in media education contest best practices. https://kcmf.or.kr/cms/index.php
14. 坂本 旬, 今度 珠美 (2018). 日本におけるデジタル・シティズンシップ教育の可能性, 生涯学習とキャリアデザイン 16(1):3–27
15. Ribble M, Bailey G (2004) Digital citizenship: focus questions for implementation. Learn Lead Tech 32(2):12–15
16. ISTE (2017) Digital citizenship education: volume 1: overview and new perspectives
17. Council of Europe (2019) Digital citizenship education handbook. https://www.coe.int/en/web/digital-citizenshipeducation/home
18. Lee JH (2017) A study on digital citizenship of social studies. J Learner-Centered Curricul Instruct 17(1):21–39
19. Common Sense Media (2009) Digital literacy and citizenship in the 21st century
20. DQ Institute (2017) Digital intelligence(DQ): a conceptual framework & methodology for teaching and measuring digital citizenship
21. Lee SY (2020) A Delphi study on digital citizenship. Seoul National University of Education Graduate School of Education. Master's thesis
22. Jenkins H (2006) Confronting the challenges of participatory culture: media education for the 21st century. In: An occasional paper on digital media and learning. John D. and Catherine T. MacArthur Foundation.
23. Hobbs R (2010) Digital and media literacy: a plan of action, a white paper on the digital and media literacy recommendations of the knight commission on the information needs of communities in a democracy. The Aspen Institute, Communications and Society Program, p 19

24. Park KB (2014) The citizenship of digital era. Korean J Element Educ 25(4):33–46
25. iKeepSafe·Harvard Berkman Center (2016) Safety, privacy, and digital citizenship
26. Rho SY (2009) Analysis of digital citizenship to implement effective digital governance. Modern Soc Public Admin 19(1):135–157
27. Hwang YS, Lee HJ, Park NS (2014) The effect of digital citizenship on on- and offline civic engagement: exploration of online civic skill. Kookmin Soc Sci Rev 25(2):493–520
28. Ahn JI, Seo YK, Kim SM (2013) The digital citizenship among adolescents—the effects of media literacy competence and the experience of education. Theory Res in Citizenship Educ 45(2):161–191

An Empirical Study on the Factors Affecting the Intent to Use and Learner Satisfaction and Education Service Platform Based on Metaverse

Hye Sun Ahn

Abstract Libraries are encountering certain challenges derived from changes on their business environment. In the case of COVID-19, digital and online functions of university libraries in Korea is suggested as an alternative for the future. Thus, it is significant to assess libraries' business efficiency focusing on digital function while it is able to acquire valuable factors for future strategy. Therefore, the study adopts DEA (Data Envelop Analysis), a widely accepted analysis model for measuring efficiency. Moreover, the study derived appropriate factors for efficiency assessment in digital service of university library by literature review. Through this, factors regarding efficiency in university libraries can be measured. We hope this study could contribute to identify inefficient factors in university libraries as well as gaining basic data for developing improvement strategy for future.

Keywords Metaverse · Education service · Self-determination · Learning flow · Flow

1 Introduction

The educational landscape has undergone significant changes due to the outbreak of COVID-19 in 2019. As a result, virtual learning environments have become increasingly crucial, and there is a growing need for non-face-to-face online education in both academic and corporate settings. A virtual learning environment refers to a system where learners can engage in educational activities through web-based platforms, allowing them to learn at their preferred locations and times while interacting with instructors. Since March 2020, most educational institutions worldwide have implemented mandatory non-face-to-face online classes. Non-face-to-face online education can be categorized into synchronous remote education, where activities occur in different times or locations, and asynchronous remote education, which

H. S. Ahn (✉)
SungKyunKwan University, Seoul, Korea
e-mail: hyesun24@skku.edu

R. Lee et al. (eds.), *AI and Metaverse*, Studies in Computational Intelligence 1160,
https://doi.org/10.1007/978-3-031-61163-6_12

involves pre-recorded content. Synchronous real-time education enables interaction during classes, often involving video conferencing between learners and instructors.

While non-face-to-face online education became a necessity due to the COVID-19 pandemic, issues such as instructors' lack of proficiency in online teaching and a decrease in the quality of education due to pre-recorded lectures have led to lower satisfaction among learners. Particularly in subjects requiring practical experience, the lack of interaction between instructors and learners has posed challenges.

Non-face-to-face education overcomes constraints related to time and location but introduces problems such as passive participation and decreased learner engagement. Moreover, although real-time video conferencing has addressed some limitations of asynchronous online education, it comes with its own set of challenges, including limited interaction and potential security concerns in video conferencing platforms. To overcome these limitations, there is a growing interest in using metaverse platforms. These platforms enable users to manipulate avatars and enhance interaction and spatial perception in virtual spaces. Initiatives utilizing metaverse for online education are on the rise. The changes in the educational environment have broader implications for future professions and educational approaches. In contemporary society, the emphasis is on fostering creative knowledge combinations and developing abilities for innovative technologies and value creation. Educational methods need to adapt to these requirements. Therefore, education should be considered comprehensively, encompassing not only school education but also corporate and initial education. Research on building new educational methods and environments using platforms like the metaverse is essential.

2 Theoretical Background

2.1 The Education Service Platform Based on the Metaverse

In recent times, the concept of the metaverse has been evolving, blurring the lines between virtual reality, augmented reality, mixed reality, and networking technologies. This integration has led to the emergence of diverse new services. For example, Second Life, which began in 2003, provides a game-like virtual world where users interact through avatars, engaging in various activities and economic transactions. However, faced with challenges such as language barriers, inappropriate content, and competition from social networking services, Second Life has seen a decline in popularity. More recently, the metaverse has regained attention, particularly among the digitally savvy MZ generation. Various metaverse-based platforms are being utilized, allowing users to enter virtual spaces, interact through avatars, and engage in high-dimensional exchanges. The realization of the metaverse requires elements such as a coherent worldview, creator economy, and everyday connections. Metaverse-based platforms offer functionalities beyond gaming, extending into various industries. The metaverse developer ecosystem is also expanding. These platforms find applications in gaming, communication, business, and are exploring potential uses in education. In South Korea, Zepeto and Gather Town are prominent metaverse platforms. Zepeto, developed by Naver, collaborates with diverse brands, demonstrating business success. Gather Town, with real-time collaboration features, is more suitable for work-related activities. Given the expansion of remote and on-tact environments due to COVID-19, the metaverse has gained momentum in South Korea, especially among the MZ generation. Human Resources Development (HRD) departments and the education sector are closely monitoring the utilization of the metaverse. Internationally, industries are actively leveraging the metaverse for job training. BMW, for instance, has established the 'Omnibus,' a virtual factory where engineers, developers, and managers worldwide collaborate to design and simulate manufacturing systems. Universities use virtual reality for medical school practical training, and companies employ the metaverse for new employee training. Walmart utilizes the metaverse for employee training, providing virtual scenarios for customer interactions and unexpected situations [1]. The metaverse platform is increasingly recognized as an alternative for education and job training, overcoming limitations of online education and enhancing learning immersion. Emphasizing the importance of experiential learning, the metaverse can provide new and immersive experiences, fostering higher levels of engagement than traditional online classes. Utilizing virtual and augmented reality in education has positive effects on learners' active cognitive processes and engagement. For instance, collaborative learning using the Second Life platform enhances interaction among learners, improving academic achievement and satisfaction. Additionally, metaverse-based virtual museums, utilizing digital content, can increase learners' motivation and interest [2]. While metaverse-based education can overcome limitations of traditional online classes and significantly enhance learning immersion, attention must be paid to relationship formation with other users and personal information protection [1]. Measures are especially necessary to prevent identity confusion and potential negative effects on adolescents.

2.2 Education Services

Education can be defined from various perspectives, and UNESCO defines it as a process that promotes individual development [3]. This process occurs through knowledge, habits, ethics, values, skills, learning, and information. Furthermore, education involves the transmission of cultural heritage from generation to generation and now encompasses critical thinking about information, essential skills for social life, vocational skills, empathy, and creative thinking.

The educational environment is primarily divided into three types:

- Formal Education: Conducted in educational and training institutions, it constitutes formal education organized according to curriculum goals and objectives. It is typically delivered by teachers or instructors, and traditional school education falls into this category.

- Informal Learning: Emerging as an additional or alternative option to formal education, informal learning takes place in flexible ways within local communities, workplaces, or civil society-based settings [4].
- Non-formal Environment: Arising from daily life experiences [5], it encompasses all educational experiences, including those within families.

Moreover, education methods include teaching, training, storytelling, discussions, directed research, and more.

The goals and values of education evolve with time, being central themes in the field of education. In modern society, education is emphasized not only for knowledge transmission but also for its crucial roles in social and economic aspects. Educational institutions are shifting their paradigms to examine education from the perspective of the learner as a customer, providing services that consider learner demands and values.

Customer-centric education refers to providing educational services as learner-centered services, prioritizing learner needs and values [6]. In this perspective, the quality of educational services reflects both the level of desired service attributes and how these attributes are perceived to be achieved. Therefore, educational services should assist learners in achieving educational goals, and providers of educational services must interact with learners in a customer-centric manner, delivering educational services in ways that satisfy the learners.

3 Autonomy

Flow, introduced by Csikszentmihalyi is a concept representing a state where action and consciousness are integrated, depicting an experience where distinctions between time, self-awareness, stimulus, and response are diminished [7]. Immersion occurs through activities such as learning and play, representing a state where the actor's consciousness and behavior are integrated, allowing for self-regulation. Csikszentmihalyi [8] highlighted various elements when explaining the state of flow [8]. It emphasizes the need for a balance between the difficulty of the task and the individual's perceived ability, clear goals, and immediate feedback. Immersion is composed of cognitive immersion and emotional immersion, and these elements are observed in learning situations. While immersion was initially central to research in leisure, sports, arts, and related fields, it has recently expanded to various disciplines, including psychology, education, sociology, and management. Immersion is a state experienced not only in learning but also in various activities, and research explores it from various perspectives, including cognitive immersion and emotional immersion.

Learning, conceptually, involves the acquisition of knowledge, skills, and experiences, triggering behavioral changes through experience. Learning is essential in everyday life, inducing behavioral changes through experience. Learning and studying can differ based on individual voluntary willingness. learning flow signifies the state where a learner experiences immersive learning activities. Learning immersion represents an optimal state where learners achieve learning goals, maximize their potential, and experience subjective satisfaction and happiness. Learning immersion stimulates interest and active participation in learning, positively impacting creativity, skill improvement, and self-esteem. Various factors, including autonomy, external motivation, learning environment, and interaction between instructors and learners, influence learning immersion. Learner immersion is influenced significantly by the learning environment and interaction. Learning immersion involves voluntary efforts for learning activities and can positively impact life quality through positive motivation and satisfaction.

4 Research Method

4.1 Research Method

While previous research has primarily focused on the successful adoption of information systems, this study explores the structural relationship between individual disposition, specifically autonomy, and its influence on the dependent variable, which is the user's intention to use, in the context of utilizing a metaverse-based education service platform.

Furthermore, the study analyzes how the quality factors of the metaverse-based education service platform, including information quality, system quality, and service quality, impact learner satisfaction through the user characteristics of utility and convenience, as well as immersion.

Finally, a model is presented to examine how learner satisfaction, perceived through the utilization of the metaverse-based education service platform, influences usage intention Fig. 1.

4.2 Operational Definition of a Variable

The study proposed a model for the success of information systems in the context of metaverse-based educational service platforms. The model combined factors from self-determination theory (autonomy and relatedness), the information system success model (information quality, system quality, service quality), and the social environment (social influence and facilitating conditions). Additionally, variables from the immersion theory and the Unified Theory of Acceptance and Use of Technology (UTAUT) were considered as mediators to understand the factors influencing platform usage intention, learner satisfaction, and the processes involved in metaverse-based educational service platform users' satisfaction and usage intention.

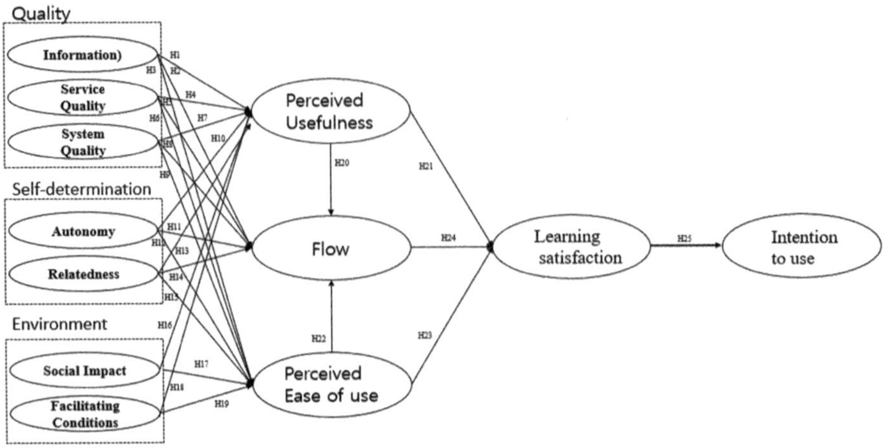

Fig. 1 Research model

The study surveyed individuals who had experienced learning at least once on metaverse-based educational service platforms. Based on prior research, the following operational definitions were provided for each factor:

(1) **Independent Variables**:

Quality: The degree to which the provided information is perceived as reliable, accurate, and up to date. It also encompasses the level of technical support provided by the service provider and the support provided by instructors.

Autonomy: The extent to which individuals feel they can act without external control or coercion.

Relatedness: The degree to which individuals perceive interaction with others and feel a sense of belonging within the learning process.

Social Influence: The extent to which individuals perceive that others believe they should use specific educational content.

Facilitating Conditions: The extent to which individuals believe there is organizational or technological support for using new technology.

(2) **Mediating Variables**:

Usefulness: The extent to which users believe their efficiency and productivity will improve in their learning activities.

Convenience: The perceived level of ease and user-friendliness when using the system.

Flow: The state in which learners are intrinsically motivated and experience happiness and enjoyment in the learning process.

Learner Satisfaction: The degree of appropriateness, efficiency, and satisfaction experienced when using the system.

(3) **Dependent Variable**:

Usage Intention: The extent to which users have a positive attitude toward activities on the platform, find them useful, and intend to use the platform again.

This comprehensive model aimed to analyze and understand the factors affecting user satisfaction and usage intention on metaverse-based educational service platforms.

4.3 Data Collection

In this study, a survey with a total of 55 items was conducted to measure users' learning experiences and factors such as quality, autonomy, environmental factors, etc., on the metaverse-based educational service platform. These items covered information quality, service quality, system quality (3 variables, 15 items), autonomy, relatedness (2 variables, 10 items), social influence, facilitating conditions (2 variables, 10 items), mediators (2 variables, 10 items), learning immersion (1 variable, 5 items), and the dependent variable (1 variable, 5 items). The survey, using a Likert scale (7 points), included general demographic characteristics such as gender, age, and usage experience. The survey data were collected from users who had experienced the metaverse-based educational service platform, resulting in a sample of 300 responses. After excluding incomplete or insincere responses (31), 269 responses were utilized for empirical analysis. A self-administered method was adopted, where respondents directly answered the survey. Statistical analysis, including frequency analysis and exploratory factor analysis, was performed using SPSS 22. Confirmatory factor analysis, internal consistency analysis, face validity analysis, and structural equation modeling were conducted using Smart PLS 3. The research model was analyzed for hypothesis testing, and group difference analysis was performed to verify moderating effects. Through these analytical processes, the study aimed to empirically validate various factors influencing user intention on the metaverse-based educational service platform.

4.4 Setting Research Hypotheses

(1) Quality Characteristics and Usability and Convenience

The quality of information is determined by how accurately users of educational service systems perceive that the learning-related information is accurate, relevant, complete, and up-to-date [9]. The quality of this information acts as a crucial factor influencing user satisfaction in the context of using education services based on the metaverse, playing an essential role in inducing usability and convenience, especially [10]. Various studies emphasize the significant impact of usability and convenience on user satisfaction during actual use, highlighting their importance in terms of usability and user satisfaction [11].

H1: Information quality will positively influence usability on the metaverse-based educational service platform.

H2: Information quality will positively influence immersion on the metaverse-based educational service platform.

H3: Information quality will positively influence convenience on the metaverse-based educational service platform.

(2) Service Quality and Usability, Immersion, Convenience

Service quality represents the comprehensive support provided by service providers to users. This support can occur through Information Systems (IS) departments, new organizational units, or outsourced internet service providers [12]. Service quality consists of four key components: responsiveness, assurance, reliability, and empathy [13].

Improved service quality contributes to enhancing user satisfaction by making services perceived as more useful and convenient. The enhancement of service quality allows users to perceive services more usefully, leading to convenience through improved services [14, 15].

H4: Service quality will positively influence usability on the metaverse-based educational service platform.

H5: Service quality will positively influence immersion on the metaverse-based educational service platform.

H6: Service quality will positively influence convenience on the metaverse-based educational service platform.

(3) System Quality and Usability, Immersion, Convenience

System quality is evaluated based on various criteria, mainly flexibility, ease of use, data quality, functionality, integration, and utility [12]. This signifies that the system is designed to be user-friendly, easy to use, and easy to connect and use by the system user [16]. System quality allows users to evaluate the system at a level that meets their expectations. Attributes such as availability, user-friendliness, ease of learning, response time, and usefulness play a crucial role in determining the quality of an online learning system [17–19].

H7: System quality will positively influence usability on the metaverse-based educational service platform.

H8: System quality will positively influence immersion on the metaverse-based educational service platform.

H9: System quality will positively influence convenience on the metaverse-based educational service platform.

(4) Autonomy and Usability and Convenience

In a metaverse-based educational service platform, learning reflects the autonomy of teachers, which originates from their desire to self-regulate their level of engagement with the chosen platform for learning [20]. This autonomy influences the perception of usability defined by the belief in the utility value related to online learning [15, 21]. To analyze how teacher autonomy impacts the learning experience, the following hypotheses are formulated:

H10: Autonomy will positively influence usability on the metaverse-based educational service platform.

H11: Autonomy will positively influence immersion on the metaverse-based educational service platform.

H12: Autonomy will positively influence convenience on the metaverse-based educational service platform.

(5) Relatedness and Usability, Immersion, Convenience

Self-determination theory includes relevance, competence, and autonomy as crucial factors increasing internal and external motivation [22]. Restraint, as one of the external factors, is described to have a positive impact on perceived usability and ease of use in sustained information systems research [23]. Furthermore, social influence, including relevance, is argued to influence individual information system use through perceived ease of use and perceived usefulness in other studies [24]. Based on this theoretical background, the following hypotheses are proposed:

H13: Relatedness will positively influence usability on the metaverse-based educational service platform.

H14: Relatedness will positively influence immersion on the metaverse-based educational service platform.

H15: Relatedness will positively influence convenience on the metaverse-based educational service platform.

(6) Environmental Influence and Usability

The relationship between social influence and technology adoption has been extensively studied in the information systems domain [25, 26]. Social influence, represented by peers influencing behavior at an individual level, such as friends, family, or colleagues, has been mentioned [27]. To investigate the impact of social influence on the use of the metaverse-based educational service platform, the following hypotheses are set:

H16: Social influence will positively influence usability on the metaverse-based educational service platform.

H17: Social influence will positively influence convenience on the metaverse-based educational service platform.

(7) Facilitating Conditions and Usability

Facilitating conditions, defined as variables directly influencing the use of new technology, represent the perception of organizational or technical infrastructure supporting the use of new technology [28]. These conditions have been noted to significantly impact learners' intention to use technology shortly after its introduction [28]. To analyze their influence on the metaverse-based educational service platform, the following hypotheses are proposed:

H18: Facilitating conditions will positively influence usability on the metaverse-based educational service platform.

H19: Facilitating conditions will positively influence convenience on the metaverse-based educational service platform.

(8) Usability

Usability and perceived usefulness are crucial factors influencing the intention to use a system in the UTAUT model [29]. Usability is defined as the belief that using a specific system [30] will enhance performance. Several studies have shown a relationship between perceived usability and the intention to accept online learning [31, 32]. To investigate their impact on the metaverse-based educational service platform, the following hypotheses are established:

H20: Usability will positively influence immersion on the metaverse-based educational service platform.

H21: Convenience will positively influence learner satisfaction on the metaverse-based educational service platform.

(9) Convenience

Convenience represents the ease with which a specific system can be used and is evaluated from various perspectives, including clarity, flexibility, ease of use, and learnability [30, 31]. Ease of use is crucial, especially for users with limited or no information communication device skills. Providing appropriate ease of use for such users can facilitate effective system utilization. To explore these relationships on the metaverse-based educational service platform, the following hypotheses are proposed:

H22: Convenience will have a positive (+) impact on learner immersion on the metaverse-based educational service platform.

H23: Convenience will positively influence learner satisfaction on the metaverse-based educational service platform.

(10) Flow

Choo-Hyang et al. [33] maintains that learning immersion, defined as a state where learners are completely absorbed in learning activities, helps maintain a psychological state of stability through the harmony between learning goals and experiences. Csikszentmihalyi (1990) identifies three conditions [8]—clear goals, the balance between challenge and skill, and immediate feedback—that promote learning immersion. From these perspectives, a hypothesis regarding how learning immersion influences learner satisfaction with the content learned in education through metaverse-based platforms can be proposed.

H24: Immersion will positively influence learner satisfaction on the metaverse-based educational service platform.

(11) Learner Satisfaction

Online learning systems, they consider the system's ability to meet their expectations [23]. Furthermore, in e-learning system use, user satisfaction significantly influences the intention to continue using the e-learning system. Therefore, user satisfaction is confirmed as a crucial factor influencing information system and online learning system adoption and continuous use.

H25: Learner satisfaction will positively influence the intention to use on the metaverse-based educational service platform.

5 Data Analysis

(1) Exploratory Factor Analysis

To validate the appropriateness of the measured variables, a factor analysis was conducted. Exploratory Factor Analysis (EFA) is a method of exploring the inherent relationships among variables without prior knowledge or predetermined assumptions [34]. Principal Component Analysis was used to extract factors, and a factor loading value of 0.4 or higher was set as the criterion. Additionally, to prevent multicollinearity, the Varimax factor rotation method with Kaiser normalization was chosen [35].

Initially, during the research, factor loadings below 0.4 were observed for seven variables among the 55 survey items. These were removed, and a secondary EFA was conducted with the remaining 48 items. The results indicated that the factor loading values for all 48 items exceeded the threshold of 0.5, confirming adherence to the criterion [36]. This process ensured that the variables were appropriately measured, enhancing the fit of the research model.

(2) Internal Consistency Reliability Evaluation

For the validation of the measurement model, concentration validity, internal consistency, and discriminant validity were applied. Internal consistency reliability was evaluated using Cronbach's Alpha, rho_A, and Composite Reliability (CR). Concentration validity was assessed using the Average Variance Extracted (AVE). Discriminant validity was evaluated applying the Fornell-Larcker criterion and Heterotrait-Monotrait ratio criterion Table 1.

Table 1 Internal consistency reliability

Latent variable	Internal consistency reliability		
	Cronbach's α	rho_A	CR
	> 0.60	> 0.70	> 0.70
Relationship	0.839	0.842	0.886
Intention to use	0.895	0.897	0.922
Flow	0.839	0.843	0.893
Social impact	0.839	0.844	0.886
Quality of service	0.767	0.771	0.866
Quality of system	0.707	0.741	0.834
Autonomy	0.811	0.817	0.869
Information quality	0.848	0.851	0.892
Convenience	0.805	0.810	0.885
Usefulness	0.828	0.830	0.897
Facilitating conditions	0.776	0.778	0.870
Learner satisfaction	0.814	0.819	0.878

(3) Validity Verification

In this study, a reflective measurement model was employed to assess the relationships between indicators of latent variables using external factor loadings and Average Variance Extracted (AVE). Concentration validity was evaluated by setting a criterion of 0.7 or higher for external loading values and 0.5 or higher for AVE values. The results demonstrated that both external loadings and AVE values exceeded the predefined criteria, ensuring concentration validity [37, 38].

Discriminant validity verification assesses whether reflective latent variables exhibit stronger relationships with their respective factors than with other factors. Two methods, following the Fornell-Larcker and Heterotrait-Monotrait ratio (HTMT) criteria, were employed for this evaluation [37].

According to the HTMT criterion, the results indicate that the correlations between latent variables meet the established criteria, ensuring discriminant validity. Furthermore, based on the Fornell-Larcker criterion, as the square root of the AVE value for each factor exceeds the highest correlation between latent variables, discriminant validity is confirmed [39, 40]. Consequently, the outcomes of this study affirm the stability and validity of the measurement tool, with discriminant validity being appropriately confirmed.

6 The Fitness of the Research Model

The fitness of the research model was assessed by analyzing the internal Variance Inflation Factor (VIF) values among latent variables. The analysis revealed that there were no VIF values exceeding 5 among the latent variables, indicating a lack of multicollinearity.

For the exogenous latent variables, an R^2 value between 0.02 and less than 0.13 indicates weak explanatory power, between 0.13 and less than 0.26 suggests a moderate level of explanatory power, and an R^2 value of 0.26 or higher signifies substantial explanatory power for the endogenous latent variables.

The adjusted R^2 value for the usefulness of the endogenous latent variable was found to be 0.696, indicating the highest level of explanatory power.

If the f^2 value for the exogenous latent variable on the endogenous latent variable is 0.02, it is considered a small effect size, 0.15 indicates a moderate effect size, and 0.35 signifies a large effect size. In the analysis of this study, the computed f^2 values indicate that learner satisfaction contributes most significantly to the path coefficient of the endogenous latent variable usage intention, with an f^2 value of 1.233, suggesting a substantial contribution.

A Q^2 value greater than 0 is indicative of predictive validity. The Q^2 values for all endogenous latent variables fall within the range of 0.383 to 0.508, affirming that the structural model demonstrates predictive validity for the endogenous latent variables. These evaluation findings support the predictive validity of the structural model for the endogenous latent variables.

The results of significance and adequacy of path coefficients, indicate that out of a total of 25 hypotheses, 7 hypotheses (H2, H5, H6, H7, H10, H12, H13) were rejected, while the remaining 18 hypotheses were accepted. Both the extended UTAUT model's mediators, usefulness, and ease of use, showed a positive and significant impact. Additionally, immersion had a positive and significant influence on learner satisfaction. Based on these results, it can be concluded that all the mediators have a positive and significant impact on the dependent variable, which is the intention to use Fig. 2.

7 Conclusions

The research results imply several key findings. Firstly, to increase the intention to use metaverse-based educational service platforms, it is necessary to adopt an integrated strategy that includes not only efficiency and functionality but also autonomy, as indicated by the study. Secondly, the research model presented in this study, which combines IS and UTAUT with an educational perspective to empirically validate general learner satisfaction and usage intention, holds significant academic importance.

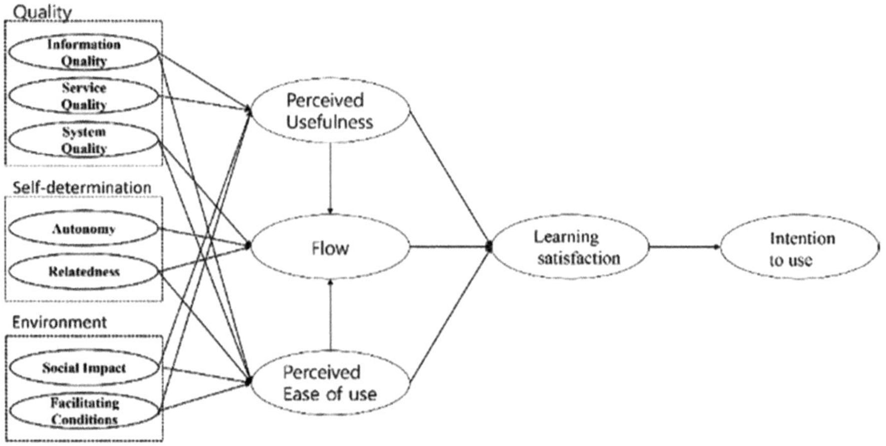

Fig. 2 Analysis results of research mode

Particularly, the outcome that immersion has a high level of influence on usage intention suggests that potential users need to recognize the advantages of immersion in metaverse-based educational services to form a positive intent to use. The study results also highlight the significance of the social atmosphere and positive evaluations from peers in platform user expansion.

Thirdly, based on the premise that personal traits such as autonomy and relatedness have a positive impact on the intention to use metaverse-based educational service platforms, the study emphasizes the need to formulate an expansive market strategy centered around individuals with high self-efficacy to transition potential users into actual users. Innovative technologies and services like the metaverse typically rely heavily on the influence and roles of early users in shaping the market. Thus, identifying the factors that impact their satisfaction is crucial for market leadership and growth.

References

1. Ki-Sung J, Seung-Hoon H, Dong-Gyu L, Jin-Mo K (2016) A study on virtual reality technology for creating immersive traditional fairy tale contents. J Comp Graph Soc 22(3):43–52
2. Mi-Hwa K (2017) Effect of performance as a team member on academic achievement and satisfaction in small group cooperative learning using virtual reality. J Digital Con 15(11)
3. Grobler S (2022) Quality education in the context of the sustainable development goals. Forum Oświatowe 34(1): 153–164
4. Singh M (2015) Global perspectives on recognizing non-formal and informal learning: why recognition matters. Springer Nature, p 220
5. Livingstone DW (2005) Expanding conception of work and learning: Recent research and policy implications. In: International handbook of educational policy, pp 977–995
6. Murdick RG, Render B, Russell RS (1990) Service operations management technical texts

7. Csikszentmihalyi M (2014) Play and intrinsic rewards. In: Flow and the foundations of positive psychology: the collected works of Mihaly Csikszentmihalyi, pp 135–153
8. Csikszentmihalyi M (1990) Literacy and intrinsic motivation. In: Daedalus, pp 115–140
9. Halonen P, Myllys L, Velmala S, Hyvärinen H (2009) Gowardia (Parmeliaceae)—a new alectorioid lichen genus with two species. The Bryologist 112(1): 138–146
10. Aparicio M, Bacao F, Oliveira T (2016) An e-learning theoretical framework 1:292–307
11. DeLone WH, McLean ER (2003) The DeLone and McLean model of information systems success: a ten-year update. J Manag Inf Syst 19(4):9–30
12. Petter S, DeLone W, McLean E (2008) Measuring information systems success: models, dimensions, measures, and interrelationships. Eur J Inf Syst 17:236–263
13. Floropoulos J, Spathis C, Halvatzis D, Tsipouridou M (2010) Measuring the success of the Greek taxation information system. Int J Inf Manage 30(1):47–56
14. Venkatesh V, Thong JY, Xu X (2012) Consumer acceptance and use of information technology: extending the unified theory of acceptance and use of technology. MIS Quart, 157–178
15. Petter S, McLean ER (2009) A meta-analytic assessment of the DeLone and McLean IS success model: an examination of IS success at the individual level. Inform Manage 46(3):159–166
16. Guimarães B, Firmino-Machado J, Tsisar S, Viana B, Pinto-Sousa M, Vieira-Marques P, Ferreira MA et al (2019) The role of anatomy computer-assisted learning on spatial abilities of medical students. Anatomical Sci Educ 12(2):138–153
17. Halawi, LA, McCarthy RV, Aronson JE (2008) An empirical investigation of knowledge management systems' success. *Journal of computer information systems*, 48.2: 121–135(2008).
18. Halawi L, McCarthy R (2007) Measuring faculty perceptions of blackboard using the technology acceptance model. Issues Inform Syst 8(2):160
19. Deci EL, Ryan RM (2013) Intrinsic motivation and self-determination in human behavior. Springer Science & Business Media
20. Pan N (2021) A comparative study on factors affecting learners' intention to continuous use of online learning. Doctoral Dissertation, Soongsil, Graduate School, Seoul, South Korea
21. Deci EL, Ryan RM (2000) The "what" and "why" of goal pursuits: Human needs and the self-determination of behavior. Psychol Inq 11(4):227–268
22. Roca JC, Gagné M (2008) Understanding e-learning continuance intention in the workplace: aself-determination theory perspective. Comput Hum Behav 24(4):1585–1604
23. Karahanna E, Straub DW, Chervany NL (1999) Information technology adoption across time: across-sectional comparison of pre-adoption and post-adoption beliefs. MIS Quart, 183–213
24. Venkatesh V, Davis FD (2000) A theoretical extension of the technology acceptance model, four longitudinal field studies. Manag Sci 46(2):186–204
25. Fulk J, Schmitz J, Steinfield CW (1990) A social influence model of technology use. Organiz Comm Tech 117(1):140
26. Irani E, Moore SE, Hickman RL, Dolansky MA, Josephson RA, Hughes JW (2019) The contribution of living arrangements, social support, and self-efficacy to self-management behaviors among individuals with heart failure: a path analysis. J Cardiovascul Nurs 34(4): 319
27. Venkatesh V, Morris MG, Davis GB, Davis FD (2003) User acceptance of information technology: toward a unified view. MIS Quart, 425–478
28. Jeon J, Jung SK (2021) Exploring the educational applicability of metaverse-based platforms. 한국정보교육학회: 학술대회논문집, 361–368
29. Doll WJ, Torkzadeh G (2021) Developing a multidimensional measure of system-use in an organizational context. Inform Manag 33(4):171–185
30. Davis FD (1989) Perceived usefulness, perceived ease of use, and user acceptance of information technology. MIS Quart, 319–340
31. Teo T (2011) Factors influencing teachers' intention to use technology: Model development and test. Comput Educ 57(4):2432–2440
32. Lin WS, Wang CH (2012) Antecedences to continued intentions of adopting e-learning system in blended learning instruction: a contingency framework based on models of information system success and task-technology fit. Comput Educ 58(1):88–99

33. Choo-hyang K, Yong-deuk J, Kwang-yong K (2018) A Study on the continuance intention of MOOC usage applying self-determination theory and learning engagement theory: an analysis of Korean and Chinese cultural differences. J Korean IT Serv Soc 17(1):121–134
34. Hoon-Young L (2015) Research methodology. Cheongram
35. Gye-Soo K (2013) SmartPLS easy structural equation. Cheongram
36. Yong AG, Pearce S (2013) A beginner's guide to factor analysis: Focusing on exploratory factor analysis. Tutor Quant Methods Psychol 9(2):79–94
37. Hair J, Hair JF, Hult GTM, Ringle CM, Sarstedt M (2021) A primer on partial least squares structural equation modeling (PLS-SEM). Sage Publications
38. Buzeta C, De Pelsmacker P, Dens N (2020) Motivations to use different social media types and their impact on consumers' online brand-related activities (COBRAs). J Interact Mark 52(1):79–98
39. Bagozzi RP, Yi Y (1988) On the evaluation of structural equation models. J Acad Mark Sci 16:74–94
40. Fornell C, Larcker DF (1981) Evaluating structural equation models with unobservable variables and measurement error. J Mark Res 18(1):39–50
41. Gefen, D., & Straub, D. W. (2000). The relative importance of perceived ease of use in IS adoption: A study of e-commerce adoption. Journal of the association for Information Systems, 1(1), 8.

Strategies for Applying AI Marketing in the Senior Market

Jungjae Lee and Jinhee Lee

Abstract The number of people aged 65 and over is expected to more than double from 700 million in 2020 to 1.5 billion in 2050. As a result, the silver industry is growing, and this silver industry should change its perception of seniors as young and active adults, not just targeting the elderly. Artificial intelligence (AI) technology, which is the most widely applied in the senior industry, is a key element in marketing strategies. Technologies such as voice recognition and the Internet of Things provide the most basic health, mobility, and convenience of life in senior life. In particular, by maximizing the convenience of seniors in the distribution and Internet markets, it is possible to present a rich life and a premium lifestyle. The expansion of the silver industry in the senior market and artificial intelligence marketing using artificial intelligence technology are closely related to each other, and strategic suggestions on how to apply artificial intelligence marketing efficiently are needed. Therefore, in this paper, we would like to examine the cognitive understanding of the senior population, the current status of each country, and the application of artificial intelligence marketing to the senior market, and present an innovative model to the changing senior market.

Keywords AI marketing · Senior · Silver industry · Artificial intelligence technology

1 Introduction

The dictionary meaning of "senior" is "olderly" person(man), "older adult", and "senior citizen". Academically, he is in the process of a complex form of interaction between physiological, physical, environmental, psychological, and behavioral changes occurring in human aging at the International Association of Gerontology in 1951. Detailed characteristics include people with defects in their own tissues who can adapt appropriately to environmental changes, people who are at a time when

J. Lee · J. Lee (✉)
Korea Soogsil Cyber University, Seoul, South Korea
e-mail: clara@mail.kcu.ac

R. Lee et al. (eds.), *AI and Metaverse*, Studies in Computational Intelligence 1160,
https://doi.org/10.1007/978-3-031-61163-6_13

their ability to integrate themselves is declining, people who are mentally deficient in their adaptation to human organs, tissues, functions, and people who are at a time when adaptation is decreasing due to the consumption of spare capacity in human tissues. In 1975, Neugarten (B. L., 1975) described the elderly in the United States in three stages. The first stage is the young-old, who is 55–64 years old, works socially, and has reached the peak of their ability to live, the second stage is middle-old, most retirees aged 65 to 74, in good health and having plenty of time to enjoy hobbies, and the third-stage elderly are over 75 years old, physically senile, economically, socially, and emotionally lonely and difficult [1]. Since then, in 1982, "elderly" means after the age of 75, and 50–75 years of age have been classified as "active seniors." Active senior refers to senior people who lead new consumption by capturing that middle-aged people with purchasing power who retired from the early 2000s began spending money for their children or families on self-improvement, fashion, and beauty [2]. These are baby boomers (born 1946–1965), who form the largest axis of the demographic structure and exert a great influence in terms of asset size and consumption. Active Senior is a rich, active, and healthy long-lived consumer group that is interested in health, enjoys consumption and leisure life, actively participates in social activities, and handles premium products and IT devices skillfully. According to recent survey data, the assets of people in their 50s and 60s range from 300 million won to 400 million won, and the proportion of people in their 50s and older using smartphones is increasing rapidly. Accordingly, it is estimated that the size of the related market will also expand. Until now, the 50s and 60s market has been limited to certain areas that are highly related to generations, such as health and insurance, but recently, interest has increased in various fields such as fashion, beauty, IT, home appliances, and healthcare, and active marketing activities are being carried out [3].

In order to enter the expanding senior market, it is necessary to understand senior consumers well. Senior consumers have several characteristics. First, trust is very important when purchasing a product. The degree is higher than that of young people, and once you build trust, it lasts longer. Second, they are conservative in trying new things. If you buy a product wrong or contract incorrectly and suffer financial losses, you will be very worried because it is difficult to make up for it. Third, they are sensitive to price and have a strong idea of reducing risk, so they are active in comparative purchases. Senior people have a lot of free time, so it is easy to collect information and have a lot of basic information because they watch the news hard. However, in addition to these characteristics, the style of purchasing goods or services differs depending on each situation, condition, and psychological state [4]. As shown in academic research, a 2009 study by Sudbury and Simcock found that senior consumers perceived their age 6–12 years younger than their actual age [5]. This reflects the desire to look younger when choosing clothes or personal items. Marketers need to closely manage senior consumers from the perspective of customer experience [4].

AI marketing is the use of AI for marketing. In other words, it is a marketing process that collects various data about customers and markets, analyzes the collected data to find marketing insights, predicts the next movement of consumers, and makes automated decisions. AI technology is applied to all of these processes, and marketers can simultaneously grasp the entire and detailed aspects of their customers through an AI-based technology platform. Using AI, you can understand consumer behavior more deeply than now. In particular, AI plays a big role in making quick judgments and decisions in an accelerated marketing environment. AI marketing strategies are important for the expanding senior market as it is necessary to better manage customer experience and apply sophisticated personalization and customized marketing activities in the senior market [6].

In particular, active seniors are taking active seniors as a new target because they are participating in various consumption activities in addition to consumption for health purposes. It is focusing on developing products and services tailored to them, such as the medical industry, distribution industry, and service industry. For example, in Japan, where the aging population is rapidly aging, new car sales are falling along with a decline in population, but the only growing type of car is sports cars. What is surprising is that interest in sports cars is increasing among people in their 50s and older. The Japanese automobile industry is meeting the demand of active seniors by diversifying sports car models in line with this trend [3]. Therefore, AI marketing application strategies are needed to more efficiently access the active senior market and satisfy their needs. In this paper, we want to understand the senior market and derive meaningful measures on how to apply AI marketing, focusing on the current status and cases of each country.

2 Theoretical Background

2.1 Conceptual Background and Research Model

The term "Old People with Active Life" refers to the economically powerful 50s and 60s, and unlike in the past, they spend generously for themselves and lead a passionate life as much as the younger generation. According to the "2021 Statistics on the Elderly," a survey conducted by the National Statistical Office, the number of senior citizens aged 65 or older in Korea is 8,537,000, accounting for 16.5% of the total population. Korea is expected to enter a super-aged society that accounts for more than 20% of the total population by 2025. Due to this change in population, various industries such as distribution and service industries have become interested, and in particular, the distribution industry has begun to pay attention to the 50s and 60s [7]. Active seniors in the aging era have attracted attention as new marketing targets in the MZ generation (born in the early 1980s and early 2000s), which was a big player in the consumer market.

Mauro F. Guillén, a professor at the Wharton School of the University of Pennsylvania in the U.S., presented his 2030 future through his book 2030: How Today's Biggest Trends Will Collide and Reshape the Future of Everything, explaining that the public's interest in millennials is exaggerated and that the generation that is actually growing rapidly in consumption is over 60 [8]. It is predicted that by 2030, the

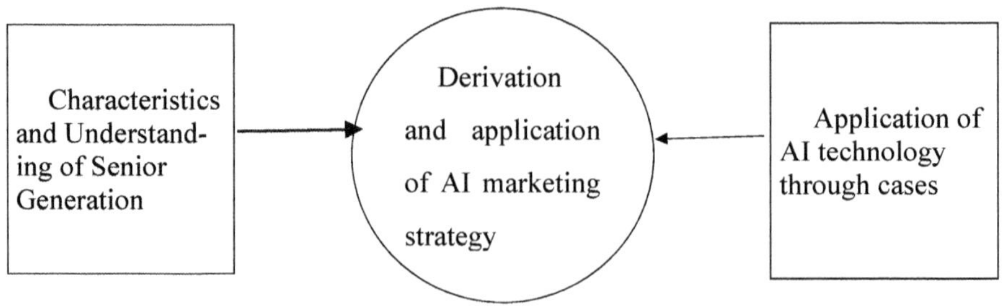

Fig. 1 Research model

number of people aged 60 and over will reach 3.5 billion worldwide. The industry predicts that senior generations in their 50s and 70s with education and assets will emerge as strong consumers in the future [9]. The competition to preoccupy them will become fiercer. Therefore, in this study, a research model was presented to examine how AI marketing should be applied, focusing on the characteristics and understanding of the senior generation and the current status and cases of each country through theoretical considerations (Fig. 1).

3 Methodology and Results

3.1 Current Status and Characteristics of Senior Generation in Different Countries

The No. 1 senior market is the U.S. The U.S. senior market estimates that it will be about \$3.5 trillion in 2025. By 2030, all 80 million baby boomers (born 1946–1965) will enter the elderly aged 65 or older. The U.S. baby boomers have more assets and pension income than their children's generation, the Millennial Generation and Generation Z, so they show high consumption power even after retirement. The senior population is emerging as a key driver in the field as a major demand group for the pharmaceutical and health care industries, and there is a high demand for advanced products and services related to medical and health care such as biotechnology, services, and regenerative medicine (Figs. 2 and 3) [10].

Japan has the highest percentage of senior population. Japan predicts that the senior market will exceed 100 trillion yen (about \$800 billion) by 2025. Japan has the highest percentage of senior population. In 2025, Japan expects the aging rate to reach 30%. The proportion of people in their 60s and older in household consumption is expected to rise steadily, reaching 47% by 2030. The consumption propensity of the elderly is increasing due to the increase in active seniors who have consumption capacity due to asset accumulation. The elderly in Japan have a home ownership rate of more than 90% and 59.5% of the total financial assets [11]. Japan's senior

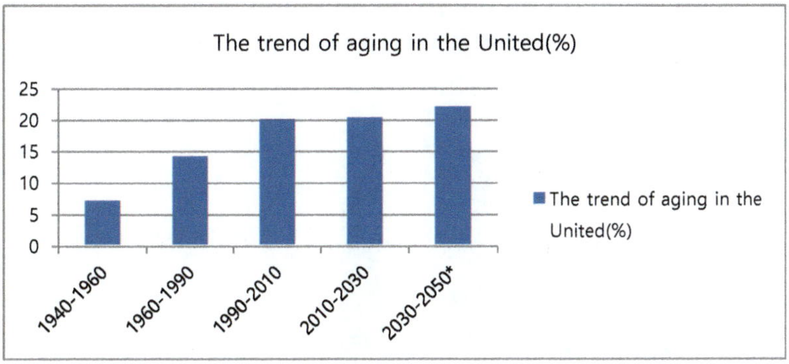

2030-2050*:estimation,Data : statista

Fig. 2 The trend of aging in the United. 2030–2050*: estimation. Data: statista

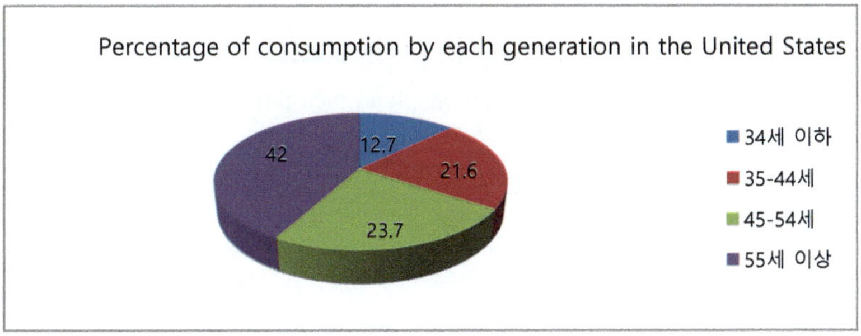

Data: KGC Direct, LCC

Fig. 3 Percentage of consumption by each generation in the United States in 2020. Data: KGC Direct, LCC

market is revitalizing products and services that reflect the needs of the elderly in various fields such as daily life, housing, medical care, and travel [12]. In particular, the household goods market is expanding its Universal Design, which is designed to be convenient for everyone regardless of the user's health or age, and the market size is so large that products considering the elderly become mainstream (Figs. 4 and 5) [11].

China also has a large senior population, and the Chinese senior market is expected to grow to 20 trillion yuan (about $3 trillion) by 2030. China had three baby boomers, and the first to third baby boomers were from 1945 to 1957, the second from 1962 to 1976, and the third from 1986 to 1990. As the population born in the second baby boom generation begins to be incorporated into the elderly, the aging population is rapidly taking place. It is an aging society with an aging rate of 7% in 2001, entered an aging society with 14% in 2021, and is expected to enter an ultra-aged society with an aging rate of 20% in 2035. The increasing number of senior citizens in China is increasing their purchasing power in the beauty, health, and fashion sectors, and the

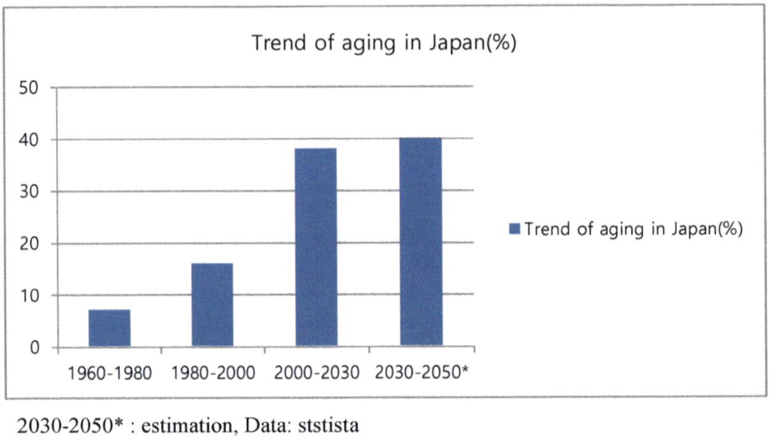

Fig. 4 Trend of aging in Japan. 2030–2050*: estimation. Data: statista

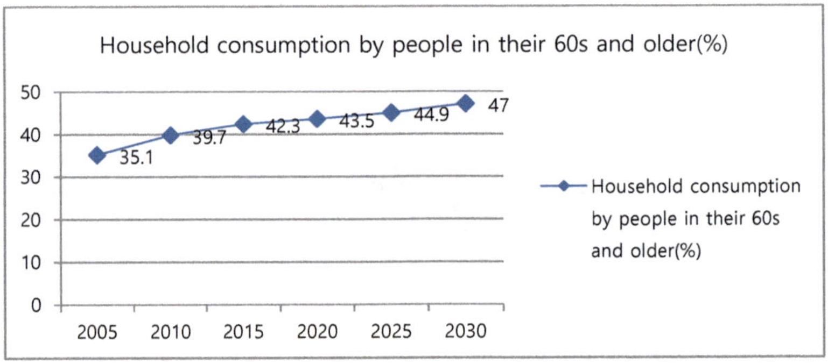

Fig. 5 Household consumption by people in their 60s and older. Data: Nisei Basic Research Institute

proportion of online shopping is increasing. The Chinese senior generation, called the "silver economy," is a generation that has begun to embrace foreign culture and has economic autonomy, and is evaluated as having strong consumption trends and potential. The purchasing power of the elderly in China is expected to expand from 26.7 trillion yuan in 2030 to 106.7 trillion yuan in 2050 (Figs. 6 and 7) [13].

In Korea, where the pace of aging is fast, the number of senior citizens aged 65 or older was 16.5% of the total population as of 2021, with 8,537,000. It entered an aged society in 2017 and is expected to enter an ultra-aged society with an elderly population ratio of 20.3% in 2025. Korea's baby boomers (1955–1963) are entering the elderly in earnest, and demand for the senior industry is expected to expand. The size of Korea's senior industry market is expected to grow from 72 trillion won in 2020 to 168 trillion won in 2030. The baby boomers are a generation that has experienced Korea's rapid economic listing and public consumption, and are

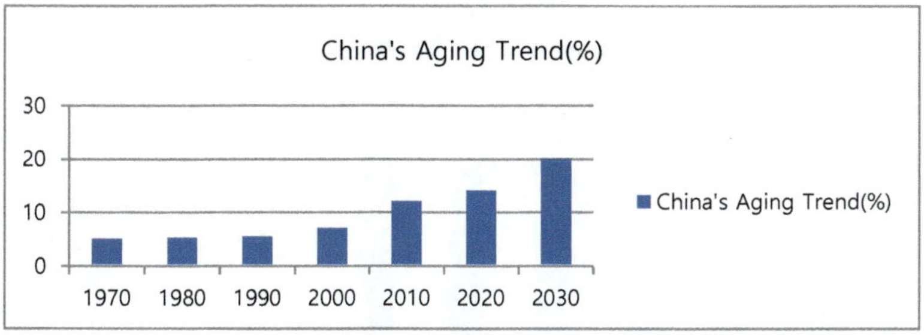

Fig. 6 China's aging trend. 2030: estimation. Data: statista

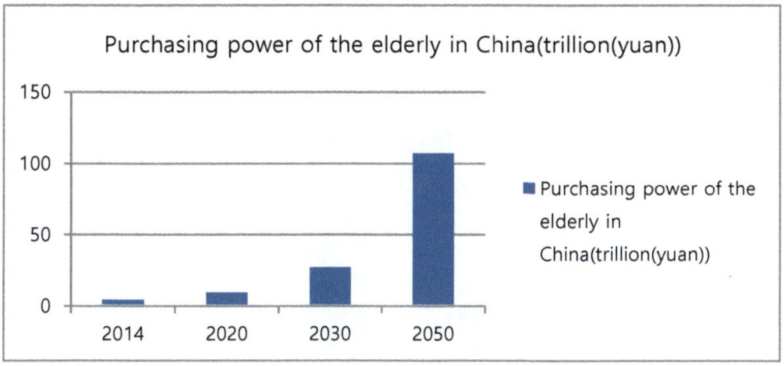

2050: estimation, Data: Chinese Research Center for Old Age Science

Fig. 7 Purchasing power of the elderly in China. 2050: estimation. Data: Chinese Research Center for Old Age Science

considered to have high cultural openness and consumption tendencies, unlike the previous generation who experienced the Japanese occupation and the Korean War. In addition, as a generation that can utilize devices and services incorporated with IT technology, the demand for IT integrated convergence devices and services is expected to be high in the future (Figs. 8 and 9) [14].

3.2 Senior Market Strategy Case

3.2.1 Understanding of Seniors

The senior generation, classified as Active Senior, has the ability to easily adapt to a new culture based on high activity and generous income. Unlike previous seniors who had been preparing for retirement depending on their children, the baby boomers are emerging as independent consumers who enjoy hobbies and consumption. When

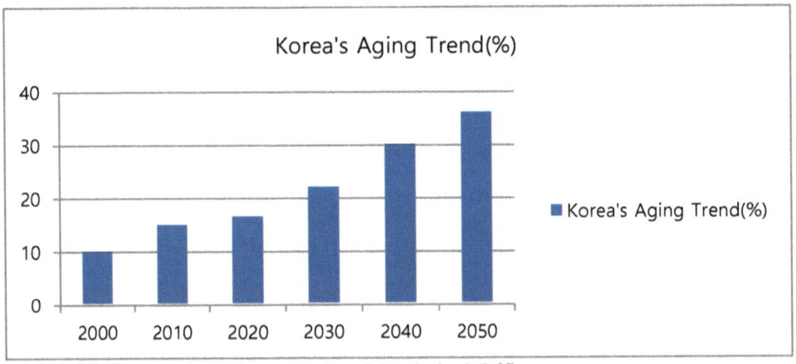

2030-2050:estimtion, Data: National Statistical Office

Fig. 8 Korea's aging trend. 2030–2050: estimation. Data: National Statistical Office

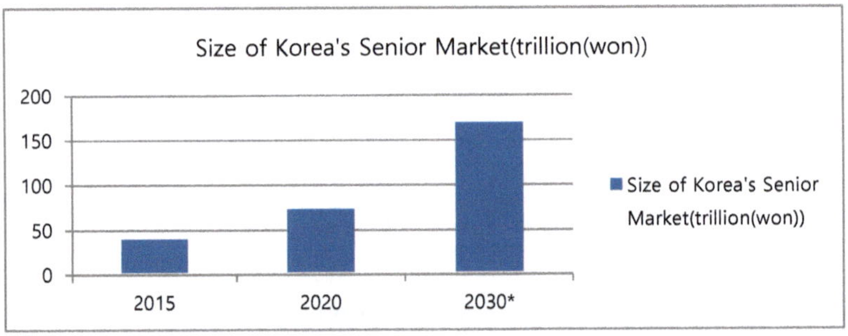

2030*: estimation, Data: Korea Health Industry Development Institute

Fig. 9 Size of Korea's senior market. 2030*: estimation. Data: Korea Health Industry Development Institute

marketing products or services, it is necessary to target active seniors by shifting the perception to young adults rather than the elderly [10]. Gold Violin, which succeeded through a change of perception, emphasized the name of the mall and the image of class, given that senior generations hate being treated as the elderly, and presented customized products that were subdivided into consideration of all age groups, not limited to the silver class. It was praised as a "stylish site" by senior members for limiting clothes, home appliances, and furniture with fresh colors and sophisticated designs to handling items [15]. Another example is the sports brand Elesse swimsuit. Usually, it comes out with a design that complements the body shape of the elderly for seniors, but Elesse is positioned as a product suitable for all ages, and uses a young design to suit the body shape of senior consumers so that it doesn't feel old. In this way, we are looking at the behavioral characteristics of seniors through AI marketing and attempting to switch to universal design that can be easily used by anyone without age distinction [10]. The "Rakuraku Phone" developed by NT DoCoMo of Japan has improved accessibility and made it easy for seniors to use

Table 1 Characteristics of active seniors

	Characteristics
A sense of life	Forward-looking
A perception of old age	Opportunity for self-realization
The attitude of life	Relaxation, enjoyment, and many hobbies
Degree of social participation	Active participation in social activities even after retirement
Retirement plan	Planned retirement
Value	Substantive and rational preference
A view of consumption	Consumption for oneself

Data: The Economist, The decode of the "young old" begins (2020)

with a simple interface on a large screen. In addition, health diary functions such as automatic heartbeat check for senior generations and built-in pedometer have been added to add simplicity and convenience, gaining popularity from all ages [16]. In Korea, there are kitchen furniture with Universal Design. For seniors who are uncomfortable to stand and suffer from back pain, Enex has designed the entire household to be removed and sit with their legs under the counter. In addition, movement was minimized by implementing a compass line that can work in a circle around the sink, countertop, and workbench. Convenience is realized by installing an automatic dish dryer with a button, a portable storage cabinet with wheels, and a shoulder height. With this convenient design, not only seniors but also wheelchair users and people with disabilities who have difficulty moving can use it conveniently (Table 1) [10].

3.2.2 Senior Emotional Stimulation

Senior generations are easily exposed to loneliness and loneliness due to bereavement with spouses, reduced exchanges with children, and reduced networks with acquaintances. Therefore, it is important to reduce social isolation by converging various auxiliary devices such as AI robots, emotional support services, and daily conversation services. Rendever, a virtual reality content developer for seniors established in 2015, started a content development project using virtual reality travel and visits to relieve the boredom and helplessness of the elderly living in silver towns and nursing homes. In 2020, it distributed products and contents to more than 100 silver towns in 39 states in the United States and expanded its business to Canada and Australia. VR content can be viewed by up to 10 people together, so it is also used as a social activity program conducted by silver towns or day care centers [17]. Recently, the contents have been expanded to health promotion programs that can be enjoyed together in welfare centers or local communities.

Intuition Robotics, a robot company established in Israel in 2015, developed a companion robot named EliQ. AI robot EliQ has developed to the level where it identifies users' characteristics through learning, expresses emotions through voice, light, and movement, and presents daily stories and light jokes first as conversation data accumulates. In addition to serving as a companion, it functions as a photo, music, phone, reading books, searching the Internet, and sending messages. These emotional exchanges can reduce social alienation [18].

3.2.3 Enhance the Convenience of Seniors

There is a high demand for services or products that will alleviate the inconvenience of daily life caused by the decline in physical function due to aging. In the late 1990s, when Japan's aging was accelerating, Keio Department Store specialized as an elderly department store targeting senior customers. 86% of department store customers are aged 50 or older, and 60% of customers are aged 65 or older. Keio Department Store is an "anti-aging" product store frequently visited by seniors and the department store placed these anti-aging products at the front and lowered the height of the store chair. In case of the skincare service corner, it has been replaced with a low-height, electric bed for senior customers. A wide elevator was installed for elderly customers in wheelchairs, and the interval between doors opening and closing was also set long. In addition, a number of chairs were placed in the middle of the store passage or in front of the elevator in consideration of the rest area [19]. Founded in 2015, Tovala developed simple food using ovens and ovens. When you partner with 20 brands and scan a barcode, it automatically cooks at the right time and intensity for the product. It is convenient for the elderly who have relatively poor eyesight and have difficulty cooking, such as setting accurate times [20].

3.2.4 Senior Care Services

The need for customized care is increasing to improve the quality of life and support independence of the senior generation. In particular, it is necessary to respond to the care needs of the senior generation, such as disease management, fall prevention, and home care support. Care Angel, which uses AI technology to monitor health conditions by voice recognition, is the first in the world to incorporate voice recognition AI into senior care services since 2016, characterizing voice answers about seniors' health conditions and whether they are taking medicine and using them as health check data. The senior's health questionnaire voice is written in the form of a report and delivered regularly to family members or doctors on a smartphone or computer. Analyzing the response, if it is recognized that it is at risk or requires special attention, an emergency alarm message is sent to your family or doctor. If the senior's health condition is not good and immediate help is deemed necessary, services such as telephone connection or virtual medical counseling services for doctor visits are provided [21]. Apple's Apple Watch Series 4 in the U.S. has added a senior feature, with accelerometers and gyroscope sensors to detect and alert users. If there is no response for more than 60s after the warning is issued, an emergency rescue request is automatically sent and a notification is sent to emergency contacts such as family members. Always monitor the electrocardiogram (ECG) and send a warning message even when the heart rate is abnormally high or low [10].

3.3 Applying AI Marketing in the Senior Market

The senior market is expanding, and overall senior consciousness and lifestyle changes are also taking place. Previously, many seniors took a defensive attitude, such as how to manage the physical decline caused by aging or how to form property and pension for retirement, but recently, an increasing number of aggressive seniors are choosing and enjoying a new life as the second stage of their lives. As the number of seniors increases, active purchases of products, motivated consumption and investment in work and hobbies are becoming more active, and related companies are also increasing their marketing opportunities. When conducting senior marketing, it is important to segment the target audience and present a product message suitable for each target. In order to lead senior marketing more successfully, it is effective to subdivide senior people by lifestyle. It is also important how to use the user experience for each lifestyle. User Experience (UX) refers to the overall experience of memories, emotions, attitudes, and actions that users feel while using a product, service, or system [22]. As the baby boomers begin retirement and emerge as a new consumer class, it is more important to design user experiences that comprehensively consider the response of the elderly in the development of products and service systems for older consumers. The combination of user experience and ICT/mobile environment has increased user behavior capabilities, and as related technologies develop, empirical knowledge can be obtained and accumulated through various analyses of users' potential needs and actions. In particular, it is important from a marketing point of view that accurate targeting can be achieved by accumulating and utilizing user experience knowledge using AI technology. Since COVID-19, many non-face-to-face ordering systems have been introduced. When I went to a fast-food restaurant and ordered hamburgers or visited a bank, the system was organized in a different way, making it inconvenient for the elderly to use it. However, by creating a system that presents menus and customized screens for the elderly in restaurants or banks, it can be developed into a stage of user experience based on anthropocentric values [23]. Previously, behavior patterns toward users were read through surveys or mechanical methods, but now marketing is applied by stacking and reflecting data through AI technology on mobile. Therefore, since the accumulation, analysis, and utilization of such user experience are important, AI marketing application is necessary to quickly grasp the changes in products and services that new seniors and seniors think about. As an example, SK Telecom's analysis of the usage patterns of artificial intelligence care services for 1,150 elderly people living alone shows that older people living alone far prefer emotional conversations compared to general users. Based on these results, when using multiple services, guidelines related to the user experience can be presented and elderly-friendly services can be implemented.

Customized content can be developed for active seniors who are emerging as new consumers. It provides customized services to induce consumption of active seniors, and Wemakeprice's 'Telemart' is an example. WeMakePrice, one of the leading e-commerce companies, introduced a service called 'Telemart'. The service was created based on the fact that older consumers find it difficult to search for information using the Internet and mobile devices. Usually, the biggest advantage of Internet shopping is its low price, but older people who are not good at searching for information do not benefit from special sales. Based on this background, the company designed a phone-on-demand sales service that creates a product catalog familiar to the elderly and deploys professional counselors to help them order and pay [24].

With the expansion of the senior market, companies are competing to produce and sell products and services accordingly, and the application of AI marketing for successful business can increase the satisfaction of senior customers by providing differentiated services based on big data.

4 Conclusions

As the number of senior citizens aged 65 or older increases rapidly around the world, the size of the silver industry is expected to continue to expand. In order to successfully enter the silver market, it is important to change the understanding and perception of seniors. It is necessary to change the perception that the senior class sees as an active entity who wants to achieve the need for self-realization, not just a passive and static object that ends life. Design designs that reflect the physical and behavioral characteristics of the elderly and products or services that enhance convenience with R&D are promising, and in particular, the design regardless of age and Universal Design designed to be comfortable for anyone can appeal to all generations. A solution that organically combines services, technologies, and products is needed to relieve loneliness or social isolation in old age, and it is necessary to provide psychological and physical health care services through empathy-based care robots, virtual reality (VR) contents, and AI-based care systems [25]. The senior market requires various segmented targeting because physical, social, and economic changes exist in various ways depending on age due to aging. Differentiated products and services that meet consumer needs should be developed through demand analysis and big data collection for not only active seniors but also those aged 65 or older. Therefore, the application of AI marketing in the senior market can meet the success of the company and the better life values and social needs of the senior class.

References

1. Kim W (2017) Cases of business innovation at home and abroad in response to senior shift in aging society, Korea Health Industry Development Institute(KHILDI), an age-friendly industry Report, 2–5
2. Lee S (2019) The influence of experiential marketing factors on attitude formation and behavioral intentions of active senior fashion consumer, Graduate School of Sangmyung University, a doctoral dissertation
3. Han S (2013) 50s and 60s of the advance, pay attention to active senior, Korea Advertisers Association, Voice of Korea Advertisers, 38–43
4. Kim J (2013) The starting point of senior marketing. LGBusiness Insight, 21–26
5. Sudbury L-R, Simcock (2009) A multivariate segmentation model of senior consumers. J Cons Market, 22–38
6. Yang K (2022) Incorporate AI into marketing, 24 June. https://www.aitimes.com/news/articleView.html?idxno=145363
7. Shin S (2019) The effects of new media utilization on the quality of life and media perception of the elderly. The Graduate School of Mass Communication Sogang University, 6–10
8. Kim M (2022) The age of aging, 5060 senior marketing active movement, 24 February. http://www.updownnews.co.kr/news/articleView.html?idxno=300489
9. Lee Y (2020) A powerful consumer group that sympathizes with all generations, move the active senior. KDI Economic Information Center, 2. https://eiec.kdi.re.kr/publish/naraList.do?fcode=00002000040000100021&sel_year=2020&sel_month=03&pp=20&pg=
10. Kim M (2022) Current status of silver market in major countries and implications for Korean enterprises. Institute for International Trade, Trade Brief, No. 12, 3–13
11. Cho E (2020) Catch Japanese silver market, Kotra, 28 May. https://dream.kotra.or.kr/user/extra/kotranews/bbs/linkView/jsp/Page.do?dataIdx=182230
12. Kim H (2004) The impact of aging population on the society and politics in Korea and Japan. Korea Social Research Center, Social Research, 81–97
13. Kwak Y (2021) 3700 trillion catch the Chinese silver industry, the main theme stock? 27 November. https://www.ajunews.com/view/20211126102718497
14. Cho H (2018) Changes in the characteristics and consumption structure of the elderly in Korea. KIET J Indust Econ Trade, 1–10
15. http://www.goldviolin.com/
16. NTT ドコモ (docomo.ne.jp)
17. Kim J (2020) U.S. start-ups catching depression in seniors with virtual reality games, 22 February. https://www.joongang.co.kr/article/23712775
18. Cho I (2017) Friends of lonely old men. AI Robot 'ElliQ', 16 January. http://www.irobotnews.com/news/articleView.html?idxno=9629
19. Kim W (2017) Keio Department Store, where 60 percent of customers are 65 or older, 24 February. https://www.mk.co.kr/premium/print/17848/
20. Chu H (2020) Smart oven startup Tobala, attraction of $20 million in investment, 3 June. https://www.digitaltoday.co.kr/news/articleView.html?idxno=236767
21. Kim J (2021) Cases and issues in caring for older adults using AI/Robots during the COVID-19 pandemic in the U.S. Inter Soc Securt Rev 16:16–25
22. Market trends and business prospects by age-friendly industry (medicine/medical device/food/cosmetic/goods) attracting attention due to aging, Impact (2023)
23. UX Design in Artificial Intelligence (2019), 22 August. https://www.wedesignx.com/knowledge/ux-in-ai
24. Active Senior Leads a New Consumption Culture (2018), 30 October. https://froma.co/acticles/248
25. Kim J (2020) Overseas cases and revitalization strategies of the silver industry. Korea Insurance Research Institute

A Study on the Impact of Quality Characteristics and Perceived Value Characteristics on the Intention to Use Mobile Electronic Certificate Services

Sae-Yong Oh and Sungtaek Lee

Abstract Certification issuance is the most frequently used e-government service by civil petitioners. However, Internet-based certification issuance services were disconnected in that users were required to have their certificate documents issued from the Internet, print them out, and submit them to relevant institutions. Most users found this extremely inconvenient. On the contrary, mobile electronic certification services enable users to experience inter-agency services through their personal mobile device. Several studies have been conducted on the acceptance of Information Systems, where the characteristics of IS Success Models were used to achieve service acceptance. Moreover, the Means-End Chain theory has indicated perceived value factors as crucial to understand the consumption behaviors of consumers. However, limited studies combine the two theories to analyze the effect of perceived value on user satisfaction and ultimately on a user's intention to use a particular service. The development of digital technology is accelerating the growth of digital transformation in our society. We hope that future studies pertaining to government innovation based on digital transformation will not only emphasize the convenience of technology but also demonstrate the importance of consumption values. Furthermore, we hope that this study contributes to the further development of e-government services and that government innovation based on digital transformation will continue to progress in our society.

Keywords Mobile electronic certificate service · IS success model · Means-end chain theory · e-government system · Digital transformation

S.-Y. Oh
SAT Information. Co., Ltd., Seoul, Republic of Korea
e-mail: ohsy@satu.co.kr

S. Lee (✉)
Department of Business Administration, Yong In University, Yongin, Gyeonggi-do, Republic of Korea
e-mail: totona22@yiu.ac.kr

R. Lee et al. (eds.), *AI and Metaverse*, Studies in Computational Intelligence 1160,
https://doi.org/10.1007/978-3-031-61163-6_14

1 Introduction

In accordance with the results of the 2020 survey on the usage patterns of the electronic government services conducted by the Ministry of the Interior and Safety (South Korea), 88.9% of the population aged 16–74 are utilizing electronic government services. Among them, 93.2% are using these services for administrative and petition document processing purposes. However, a drawback was identified in situations where individuals had to submit petition documents to relevant agencies; they were required to print and submit the documents obtained online, leading to considerable inconvenience for service users. Therefore, this study identifies the necessity for the following research through a review of literature.

Firstly, the innovation of services due to digital transformation instills expectations for change in consumers. Consumer expectations play a crucial role in service acceptance, manifesting clear results in satisfaction and dissatisfaction with services. Therefore, this study contends that an analysis of quality factors influencing the intention to use mobile electronic certificate services is required through the Information Systems Success Model.

Secondly, according to the Means-End Chain theory, which integrates the Information Systems Success Model and the value of consumerism, consumers attribute decisive roles to values before the psychological states of satisfaction or dissatisfaction in consumption behavior. The theory suggests that consumers simplify the selection process by connecting their values to their consumption behavior [1].

Therefore, this study aims to analyze the impact of quality factors of mobile electronic certificate services on users' perceived value and satisfaction, ultimately examining their influence on service intention to use by combining the Information Systems Success Model and the Means-End Chain theory.

2 Theoretical Background

2.1 Mobile Electronic Certificate Service

The mobile electronic certificate service allows individuals to submit various certificates directly to relevant institutions or submission points using their smartphones. To facilitate this, the mobile electronic certificate service is a system that issues various administrative documents, such as certificates, in the form of electronic certificates and enables their electronic distribution through the use of a personal electronic document wallet [2]. Users can apply for the issuance or reissuance of electronic certificates through the apps of selected institutions (government, public, financial institutions, etc.) and the service is provided through the Government 24 platform.

2.2 *Characteristics of Mobile Electronic Certificate Services*

2.2.1 System Quality

System Quality refers to the overall quality of a system as perceived by users while utilizing the information system that generates information.

Wang et al. [3] emphasized the importance of stable system quality factors for online services as a critical success factor for mobile e-government. Heo and Seong [4] defined system quality for e-government as the clarity and reliability of the relevant business system, as well as the responsiveness and information delivery capability of the web page.

In this study, drawing from previous research, system quality is presented as the convenience of access to mobile electronic certificate services, the user-friendly configuration provided by the system, and the seamless use of communication devices by users.

2.2.2 Reliability

Reliability refers to the consumer's trust and expectations regarding the system and services. Zeithaml [5] defined reliability as the certainty of the functions of the site and the information provided about the products and services. Parasuraman et al. [6] defined reliability as the ability to trust the service and perform it accurately. The belief that the e-government service will operate responsibly is crucial for the success of e-government services when users visit or transact with administrative services.

In this study, drawing from previous research, reliability is presented as the accurate performance ability of mobile electronic certificate services in carrying out tasks and the user's trust in the service.

2.2.3 Accessibility

Accessibility refers to the extent to which users can access the service anytime, anywhere in the context of service utilization. Zeithaml [5] defined accessibility as the ability of customers to access the site whenever they desire. Accessibility is a crucial factor determining the success of information systems and is a key metric used in measuring system adoption.

Li and Shang [7] analyzed the success factors of e-government services, presenting accessibility, reliability, information quality, interactivity, and security as service quality factors. They found that the accessibility of the system significantly influences citizens' intention to adopt services.

In this study, drawing from previous research, accessibility is presented as the degree to which users can access mobile electronic certificate services without constraints of time and place, contributing to user convenience.

2.2.4 Information Quality

Information Quality refers to the measurement of whether the information provided by an information system is suitable for the consumer's intended purpose.

Wang and Teo [8] stated that an environment where citizens using online services in mobile e-government can obtain prompt and accurate information is crucial for the success of mobile e-government. Kumar [9] mentioned that information quality and trust in services influence the continuous intention to use e-government services. Kim [10] defined information quality as the degree to which an e-government website provides information promptly and accurately to users, suggesting that it can have a positive impact on citizen satisfaction and intention to use.

In this study, drawing from previous research, information quality is presented as the user's perception that mobile electronic certificate services will deliver accurate and relevant information, ensuring the currency of the information.

2.2.5 Service Capability

Service Capability refers to the service provider's ability to offer one-stop services to users utilizing the service. Tan et al. [11] and Hu et al. [12] indicated that service capability represents the content provision capability of government websites to meet citizens' demands.

Li and Shang [7] emphasized the importance of service capability as a service quality factor for citizens' sustained intention to use e-government services, analyzing that it plays a positive role in the continuous intention to use government services through satisfaction and trust in the government. Choi [13], in a study on the impact of quality factors of e-government services on service satisfaction, identified that accessibility, interactivity, responsiveness, and service capability have a positive impact on service satisfaction.

In this study, drawing from previous research, service capability is presented as the convenience for users of mobile electronic certificate services to use their mobile devices through a single channel for the issuance and submission of certificates.

2.2.6 Interactivity

Interactivity refers to the actions performed by individuals in a given environment. Lombard and Snyder-Duch [14] defined interactivity as all the actions exchanged between humans and objects or humans and existing entities.

In a study on the revisit relationship of websites, Dholakia et al. [15] presented user controllability, real-time interaction, responsiveness, connectivity, and enjoyment as characteristics of interactivity.

In this study, drawing from previous research, interactivity is presented as the belief that information sharing between the service provider (institution) and the user will interact through mobile electronic certificate services.

2.2.7 Responsiveness

Responsiveness refers to providing services to consumers promptly. Parasuraman et al. [6] defined responsiveness as the attitude of providing services promptly to customers. Trocchia and Janda [16] presented responsiveness as the business capability that instills trust in consumers, meaning the prompt handling of customer demands, including immediate service processing speed.

Li and Shang [7] identified responsiveness as a service quality factor for e-government services, suggesting that quick responses to service failures, such as text notifications and email alerts, provide trust in the government to users and encourage continued service usage.

In this study, drawing from previous research, responsiveness is presented as the ability to promptly address issues arising while using mobile electronic certificate services and the capability to handle user requirements swiftly.

2.2.8 Perceived Value

Perceived Value refers to the perceived benefits or psychological prices obtained through the sacrifices of the consideration paid by consumers. Generally, it can be defined as the perceived evaluation through the difference between the consideration sacrificed for the act of consumption and the benefits gained by the consumer [17].

In this study presents perceived value as the value experienced by users while utilizing mobile electronic certificate services. The issuance of certificates is one of the most in-demand services in e-government services. It highlights the factors contributing to the value offered through innovative services facilitated by digital transformation. This includes maximizing user convenience as an efficient value factor, public value in terms of a transparent and responsible government perceived by consumers while using government-provided services, and inclusive value as a service catering to information-vulnerable demographics.

2.2.9 User Satisfaction

User satisfaction is defined as the level of satisfaction experienced by users while utilizing mobile electronic certificate services. Satisfaction involves maintaining a positive attitude toward the service, increasing customer loyalty, and leading to repeated service usage.

Wang et al. [3] emphasized the impact of system quality, information quality, and online service quality as success factors for mobile-based e-government services on citizen satisfaction, enhancing the value of service usage and influencing the success of government services. Li and Shang [7] indicated that quality factors of e-government services influence the improvement of users' perceived value, which, in turn, affects user satisfaction and the intention to use the service.

In this study defines satisfaction as user satisfaction when utilizing mobile electronic certificate services, emphasizing that the service quality of mobile electronic certificate services contributes to user satisfaction and meeting user expectations.

2.2.10 Intention to Use

Intention to use refers to the subjective likelihood that beliefs and attitudes towards a product or service will be translated into behavior. It serves as an intermediate variable between an individual's attitudes and behaviors [18], representing the subjective probability of a consumer's future actions based on the transformation of belief attitudes into behaviors [19].

The success of mobile electronic certificate services is determined not only by technical factors but also by the intention of the public, the service users. Therefore, analyzing the intention to use is crucial as it directly influences the decision factors for service usage.

In this research, referring to previous studies, 'intention to use' is defined as the willingness of users to actively engage in using mobile electronic certificate services. This includes prioritizing the use of these services, actively recommending them to others, and expressing a positive willingness to engage in service usage.

2.3 *Literature Review on Research Models*

2.3.1 Information System Success Model

The quality factors of information systems, including system quality, information quality, and service quality, have been identified as factors influencing service users' intention to use, use, user satisfaction, and the overall effectiveness of information systems. This enhancement aims to improve the information system success model, a theory widely used for evaluating the quality of information systems across various fields [20].

Wang and Teo [8] presented mobility, location confirmation, personalization, and security as service quality factors for mobile e-government services. They highlighted the significant impact of these quality factors on user satisfaction and the intention to use the services. Li and Shang [7] emphasized the influence of service quality and trust in the government on user satisfaction, consumer loyalty, and usage behavior in e-government services.

In this study, we drew insights from the information system success model and numerous previous studies related to the intention to use e-government services, allowing us to analyze the impact of quality factors in mobile electronic certificate services on the intention to use the services.

2.3.2 Means-End Chain Theory

The Means-End Chain theory is one of the cognitive models applied in consumer behavior research to understand the motivations behind product or brand selection [21]. According to the Means-End Chain theory, values play a crucial role in triggering consumer behavior, and consumers simplify the decision-making process by associating their values with consumption actions [1]. In essence, the attributes of a product or brand serve as means to achieve desired outcomes (ends) in the consumer's perspective. Attributes, in this context, are not inherently meaningful but influence higher-order concepts like life values more than the positive or negative results they deliver to the consumer [1].

The most critical concept in the Means-End Chain theory is Values, which signify the ultimate objectives individuals pursue in life, such as happiness, safety, and success. Since these objectives are subjectively and internally felt by individuals, the outcomes are positioned below values in the hierarchical structure [1].

3 Research Model and Hypothesis

3.1 Research Model

In this study, we propose a research model that combines the Information System Success Model and the Means-End Chain theory to analyze the impact of the quality characteristics of mobile electronic certificate services on intention to use. The Information System Success Model provides a theoretical framework for understanding the influence of service quality characteristics on user satisfaction and intention to use, while the Means-End Chain theory explores the impact of service quality characteristics on perceived value and intention to use.

To achieve this, we build on previous research and adopt variables such as Fig. 1.

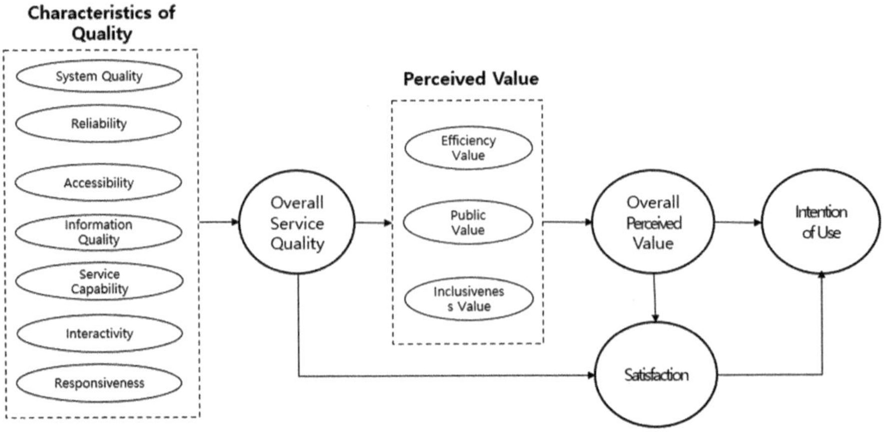

Fig. 1 Research model

3.2 Hypothesis

H1 : The satisfaction(SAT) with mobile electronic certificate services is expected to have a positive (+) effect on intentions to use(INU).

H2 : The overall service quality(OSQ) of mobile electronic certificate services is anticipated to have a positive (+) impact on satisfaction(SAT).

H3 : The overall service quality(OSQ) of mobile electronic certificate services is expected to have a positive (+) impact on overall perceived value(OSV).

H4 : The overall perceived value(OSV) of mobile electronic certificate services is predicted to have a positive (+) effect on satisfaction(SAT).

H5 : The overall perceived value(OSV) of mobile electronic certificate services is expected to have a positive (+) effect on intention to use(INU).

4 Research Model and Hypothesis

4.1 Sample Characteristics

Analysis included responses from 364 participants, consisting of 241 males (66.2%) and 123 females (33.8%). In terms of age distribution, there were 5 respondents in their 10s (1.4%), 25 in their 20s (6.9%), 46 in their 30s (12.6%), 129 in their

40s (35.4%), 146 in their 50s (40.1%), and 13 aged 60 or above (3.6%). Regarding previous experience with certificate issuance, 339 participants (93.2%) had prior experience, showing a significant difference compared to the 25 participants (6.8%) without experience.

4.2 Hypothesis Testing

The Cronbach's Alpha values for each variable were all above 0.7, indicating that the reliability of all 12 variables has been secured. The evaluation results for the significance and appropriateness of the hypothesis paths in the proposed research model are presented in Table 1. According to the evaluation, all five hypotheses were accepted.

The results of hypothesis testing for the research model are presented in Table 2. When the t-value is $t \geq 1.96$, statistically significant differences between the two groups are observed at a significance level of $p < 0.05$ [22].

The verification of standardized coefficients shows that all five hypotheses related to the intention to use mobile electronic certificate services have been accepted. The p-values for all five hypotheses are smaller than the permissible range of 0.05.

According to Table 1, the results of hypothesis testing for the positive impact on the intention to use mobile electronic certificate services are as follows:

Table 1 Hypothesis testing

Paths		Original sample (O)	T statistics (|O/STDEV|)	*P* values	Accepted/rejected
H1	SAT → INU	0.414	5.833	0.000	Accepted
H2	OSQ → SAT	0.523	7.739	0.000	Accepted
H3	OSQ → OSV	0.788	29.117	0.000	Accepted
H4	OSV → SAT	0.328	5.126	0.000	Accepted
H5	OSV → INU	0.503	7.29	0.000	Accepted

Table 2 Specific indirect effect in the research model

Paths	Original sample (O)	Sample mean (M)	Standard deviation (STDEV)	T statistics (|O/STDEV|)	*P* values
OSQ → OSV → INU	0.376	0.374	0.058	6.530	0.000
OSQ → OSV → SAT → INU	0.107	0.109	0.028	3.891	0.000
OSQ → SAT → INU	0.238	0.240	0.054	4.420	0.000
OSV → SAT → INU	0.136	0.138	0.034	3.965	0.000
OSQ → OSV → SAT	0.241	0.244	0.049	4.893	0.000

H1: The path relationship between Satisfaction (SAT) and Intention to Use (INU) is positive and significant at the significance level p = 0.

H2: The path relationship between Overall Service Quality (OSQ) and Satisfaction (SAT) is positive and significant at the significance level p = 0.

H3: The path relationship between Overall Service Quality (OSQ) and Overall Perceived Value (OSV) is positive and significant at the significance level p = 0.

H4: The path relationship between Overall Perceived Value (OSV) and Satisfaction (SAT) is positive and significant at the significance level p = 0.

H5: The path relationship between Overall Perceived Value (OSV) and Intention to Use (INU) is positive and significant at the significance level p = 0.

4.3 Indirect Effect

According to the p-values presented in Table 2, it can be concluded that the mediating effect of OSV on the relationship between OSQ and INU is significant. Additionally, the mediating effects of OSV on the relationship between OSQ and SAT, the relationship between OSQ and INU, the relationship between SAT and INU through OSV, and the mediating effect of OSV on the relationship between OSQ and SAT are all deemed significant.

5 Result

In the research model, five hypotheses were formulated, and ultimately, all of them were accepted. In summary, overall service quality characteristics had a positive impact on both overall perceived value and user intention. Seven variables representing sub-factors of overall service quality characteristics were utilized, encompassing attributes related to system quality, information quality, and service quality. Consistent with prior research, the quality characteristics of mobile electronic certificate services were found to positively influence users' perceived value. Furthermore, overall service quality characteristics positively influenced user satisfaction and intention. As mobile services have become integrated into daily life, the convenience they provide may be perceived not only as satisfying but also as a dissatisfaction if not supported by mobile services. Consumers seem to derive satisfaction from mobile electronic certificate services, which, in turn, positively influences their intention to use the service. Lastly, overall perceived value was observed to have a positive impact on both user satisfaction and intention. In the research model, latent variables represented the sub-factors of perceived value, while overall perceived value served as a measured variable for analysis. The results indicate that the quality characteristics of mobile electronic certificate services have a positive influence on users' perceived value, satisfaction, and intention. This underscores the significance of the combined theoretical framework of the Information System Success Model and Means-End Chain theory in providing meaningful results in this study.

References

1. Gutman J (1982) A means-end chain model based on consumer categorization processes. J Market, 60–72
2. Ministry of the Interior and Safety (2019) Request for proposal for the establishment of an electronic certificate issuance and distribution system
3. Wang C, Teo TSH, Liu L (2021) Perceived value and continuance intention in mobile government service in China. Telematics Inform, 48
4. Heo D, Seong W (2020) A study on the impact of quality, intention to use, and satisfaction of e-government systems on work performance: focusing on government 24 (Civil Service Portal). J Korean Reg Inform Soc 23(1):25–50
5. Zeithaml Valarie A, Zeithaml Valarie A (1988) Consumer perceptions of price, quality, and value: a means-end model and synthesis of evidence. J Mark 52(3):2–22
6. Parasuraman A, Zeithaml VA, Berry LL (1988) SERVQUAL: a multi-item scale for measuring consumer perception for service quality. J Retail 64(1):12–37
7. Li Y, Shang H (2020) Service quality, perceived value, and citizens' continuous-use intention regarding e-government: Empirical evidence from China. Inform Manage 57(3)
8. Wang C, Teo TSH (2020) Online service quality and perceived value in mobile government success: an empirical study of mobile police in China. Int J Inform Manage, 52
9. Sharma SK, Al-Badi A, Rana NP, Al-Azizi L (2018) Mobile applications in government services (mG-App) from user's perspectives: a predictive modelling approach. Gov Inf Q 35(4):557–568
10. Kim S (2019) A study on the trust in e-government websites and citizens' satisfaction and continuous use intention: focusing on trust and information system success model. Korean J Local Gov Admin Stud 33(2):91–116
11. Tan C, Benbasat I, Cenfetelli RT (2013) IT-mediated customer service content and delivery in electronic governments: an empirical investigation of the antecedents of service quality. Mis Q 37(1):77–109
12. Hu G, Pan W, Lin H, Kang K, Best ML (2014) Study on the framework of e-government services capability: an empirical investigation. Soc Sci Comput Rev 32(1):56–73
13. Choi Y (2018) A study on the impact of e-government service types and quality on service satisfaction: focusing on the moderating effect of service engagement. J Korean Gov 25(3):145–172
14. Lombard M, Snyder-Duch J (2001) Interactive advertising and presence: a framework. J Interact Advert 1(2):56–65
15. Dholakia LL, Zhao M, Dholkia N, Fortin DR (2000) Interactivity and revisits to websites: a theoretical framework. RITIM Working Paper
16. Trocchia PJ, Janda S (2003) How do consumers evaluate internet retail service quality? J Serv Mark 17(3):243–253
17. Gallaeza MG, Saura IG (2006) Value dimensions, perceived value, satisfaction and loyalty: an investigation of university students' travel behavior. Tour Manag 27:437–452
18. Engel JF, Blackwell RD, Paul WM (1995) Consumer behavior. The Dryden Press, New York
19. Aaker DA (1996) Measuring brand equity across products and markets. Calif Manage Rev 38(3):102–120

20. Rai et al (2002) Assessing the validity of is success models: an empirical test and theoretical analysis. Inform Syst Res 13(1):50–69
21. Cho H (2016) Means-end chain model: a review of domestic research, 2001–2015. J New Ind Bus 34(1):1–25
22. Henseler J, Ringle CM, Sinkovics RR (2009) The use of partial least squares path modeling in international marketing. In: New challenges to international marketing. Emerald Group Publishing Limited

A Study on the Professional Baseball Spectating Factors Affecting the Continuance Intention Focus on Dynamic Pricing

Jae-Mo Lee

Abstract Dynamic pricing refers to a pricing strategy that flexibly changes the price of the same product or service according to market conditions as it becomes possible to reproduce the price control function through an invisible hand of AI and big data. Using real-time and interactive technology, it is possible to meet the changing price needs of consumers and satisfy the complex needs as well. In this study, based on the IS Success Model presented by Delone and Mclean' study and the expectation confirmation model by Bhattacherjee' study, NC Dinos' AI and big data applied dynamic pricing were used to set the viewing factors for professional baseball. The purpose of this study is to identify the factors affecting the continuous intention to use. While previous studies mainly focused on identifying the factors that reflected the value of professional baseball, this study is meaningful in that it empirically analyzed the quality of the system in relation to dynamic pricing.

Keywords Dynamic pricing · Professional baseball spectator · Price fairness · Satisfaction · Continuance intention

1 Introduction

Industrialization and economic growth led to an improvement in people's standard of living and an increase in leisure time. Sports have become a major interest and an important industrial market as part of leisure life. While the demand for spectator and participatory sports has increased, professional sports have continued to grow since their introduction. In particular, baseball became the first professional sport among domestic sports.

The annual cumulative number of spectators in domestic professional baseball reached a peak of 8,400,688 in 2017 and then turned to a downward trend. Afterward, due to COVID-19 pandemic, social distancing adjustments were strengthened

---

J.-M. Lee (✉)
Department of School of Business Administration, Soongsil University, Seoul, South Korea
e-mail: jaemo.lee@outlook.com

R. Lee et al. (eds.), *AI and Metaverse*, Studies in Computational Intelligence 1160,
https://doi.org/10.1007/978-3-031-61163-6_15

in accordance with the government's quarantine guidelines, so professional baseball played games without spectators, etc. 328,317 in 2020 1,228,489 in 2021, and 6,076,074 in 2022.

Mobilization of spectators in professional sports not only affects competitiveness but is also the most basic source of income for clubs [1]. The increase in the number of spectators is directly related to not only admission revenue, but also sponsorship, broadcasting rights fees, and product sales revenue, which are the sources of derived income for professional teams. However, since the number of spectators and the size of the stadium are finite, an appropriate ticket price is an important factor to consider in order to increase the club's profits. A number of industries have used dynamic pricing strategies to maximize profits within a limited set of customers. A typical example of a revenue management strategy in the air transportation industry is the operation of a dynamic pricing strategy for the same seat [2].

The professional sports industry also introduces a strategy to increase revenue by changing the price of services based on expected demand. A number of professional sports clubs have used a strategy to increase ticket prices, such as on weekends when spectator demand is expected to be high [3].

The regular season in professional baseball generally starts in early April and ends in late October and lasts a relatively long period compared to other sports [4]. According to a study by Lemke et al. on the factors affecting game spectator demand in the MLB season, except for the top teams, the number of spectators decreased in September [5].

2 Theoretical Background

2.1 Dynamic Pricing

Dynamic pricing refers to a pricing strategy that flexibly changes the price of the same product or service according to market conditions as it becomes possible to reproduce the price control function, which is known as an invisible hand, using AI and big data. In other words, it refers to dynamic pricing, which regulates demand by raising prices when demand increases and lowering prices when demand decreases. It was defined as a pricing strategy in which price is useful for customers, time, products, or services [6]. It implements and develops accurate pricing strategies in response to market and competition conditions [7].

The advantage of price flexibility is that consumers can choose the purchase price. However, due to the continuation of price fluctuations, consumers constantly predict the price they will purchase and repeat this process until the purchase decision is completed.

In the past, even if the price had liquidity, only the fixed price and the current price were presented, making it easy for consumers to use information heuristics. However, due to price anchoring, dynamic pricing causes consumers to become aware of other consumers who purchase at a lower price than they do, thereby stressing their purchase. Through this, consumers continue to process information such as self-justification for their decisions even after purchase.

In fact, we investigated whether these characteristics of dynamic pricing affect consumers' decision-making and information processing. It was found that consumers repeat the process of controlling their own pricing in the process of participating in pricing, and come to have the idea that they purchased the purchased price at the desired price. This means that in the dynamic pricing purchasing process, the buyer has an illusion of control over the transaction and price [8].

It may seem unfamiliar that prices are flexible, but this flexible pricing method is actually a familiar pricing strategy that has existed for a long time. Dynamic pricing started to attract attention because the fixed price system in the era of mass production did not fulfill its function in the new economic order driven by the Internet, such as B2B and B2C. However, dynamic pricing can satisfy the ever-changing price needs of consumers and satisfy complex consumer needs by using interactive technology in real-time.

A representative example is Priceline, which provides travel-related online reservation services such as hotels, airlines, and car rentals. Established in 1998, Priceline introduced NYOP (Name Your Own Price), a new pricing strategy based on a reverse auction method in which suppliers compete with each other to determine the final price when consumers present the amount they are willing to pay. Through this service, consumers could obtain the same product at a readjusted price according to competition between suppliers or the product's inventory status. As this strategy proved successful, other online retailers began introducing their own dynamic pricing. Amazon is known to have increased sales by about 27% compared to before the introduction of dynamic pricing in 2012.

The digitalization megatrend that has recently appeared throughout the entire industry is creating a more favorable environment for the spread of dynamic pricing. The development of various ICT technologies, such as cloud computing, big data analysis, and artificial intelligence algorithms, has further advanced dynamic pricing, enabling personalized pricing optimized for individual consumers in consideration of their circumstances and tastes, as well as simple supply and demand. Dynamic pricing, which has become so powerful, is appearing in various forms across a variety of industries. The development of ICT technologies such as IoT and AI continues, and as digitalization becomes deeply embedded in our lives, the area of application and influence is expanding.

Dynamic pricing is more efficient than static pricing because it evaluates the value felt by each consumer more accurately and reflects it in the price. In this respect, the core of dynamic pricing can be said to be in the evaluation of consumers' perceived value. Each person has different values, and even if the same person receives the same product or service, the value experienced may vary depending on the surrounding circumstances, such as the timing and location of the service. The ability to understand the dynamic complexity of these consumers and connect them to reasonable pricing is a key competency essential for the success of dynamic pricing.

2.2 Price Fairness

According to Adams' study, when comparing the economic results of one's own input with the economic results of other people's input, it is fair when the ratio is balanced [9]. Fairness is an important concept in the literature of social psychology and organizational psychology, and the importance of equity theory in the field of marketing has been asserted by many studies.

Price fairness, which originated from equity theory, is known as a very important factor in consumer response related to price. Price fairness is the buyer's perception of the fairness of the seller's price [10]. Price fairness is defined as judging whether the difference between the seller's price and the reference transaction is acceptable, justifiable, and reasonable [11]. Price is a cost to the buyer and a source of profit to the seller. Consumers have standards for reference transactions that they believe will benefit both buyers and sellers, and if the profit-loss ratio of reference transactions that consumers believe breaks due to the difference in price, they may feel price unfairness. According to Kahneman et al.'s study, the principle of dual entitlement was presented, which means that the seller and the buyer have the right to the reference profit and the right to the reference price, and the selling price determined according to the company's reference profit When compared with the past purchase experience or the price of a competing product, a person perceives price unfairness if they feel that the reference price is infringing [12].

Unfairness can be perceived more in the case of loss than in the case of gain. Also, even more in the case of the reference transaction. Consumers want to maximize their performance and try to judge price fairness based on their own profit [13]. Consumers consider the difference in comparison with the reference transaction to be fair when it is advantageous but judge it to be unfair if it is considered unfavorable to them. In other words, consumers do not express negative feelings about advantage inequity. Consumers' perception of price fairness can be viewed very subjectively. Consumers tend to ignore the effect of price increases in the difference between the seller's current selling price and the previous price, and even if a company raises prices, they perceive it fairly if it is related to cost increases, and the cause of the price increase is the company's If it is judged to increase profits, the perception of fairness is lowered [11].

In other words, the reference price perceived by consumers is an axis constituting the principle of bilateral authority, which affects consumers' perception of price fairness, and perceives unfairness when the difference between the reference price and selling price is a loss compared to a gain. The perception of fairness or unfairness in price has an impact on consumer satisfaction, purchase intention, and complaining behavior [14, 15].

2.3 Information System Success

Delone and Mclean analyzed 180 papers on the success of information systems published from 1981 to 1987. These variables are related to the success of information systems and are organized into six categories: system quality, information quality, use, user satisfaction, individual impact, and organizational impact, and a new model was proposed by defining the relationship between each variable, which served as the basis for future research [16].

As seen in (Fig. 1), the variables of the information system success model are in a relationship that affects each other. System quality and information quality affect use and user satisfaction, and at this time, use and user satisfaction affect each other. This affects the individual impact, and the individual impact affects the organizational impact. The variables of the information system success model are correlated with each other in an interdependent manner.

This was done by collecting and arranging existing studies on Delone and Mclean's information system success model research and classifying items that measure success into appropriate groups. It was meaningful in that it provided a theoretical and empirical basis for the information system success model [17]. However, Delone and Mclean also said that more research is needed to derive the items of an appropriate success model because this model is still incomplete and unclear.

Pitt et al.'s study said that the study of Delone and Mclean focused only on system quality and information quality and overlooked the aspects of information systems [18]. This was due to the fact that the study was based on data from the mainframe era while analyzing research in the early 1980s. Since the overall effectiveness cannot be measured if the service aspect is excluded from the information system success model, it is necessary to measure the performance of the quality of the service aspect. Delone and Mclean's Model of IS Success.

Delone and Mclean presented an updated model after the existing information system success model [19]. The difference from the existing ones is that the service quality proposed by Pitt et al.'s study is added, and individual impact and organizational impact are integrated into a single variable called net benefits.

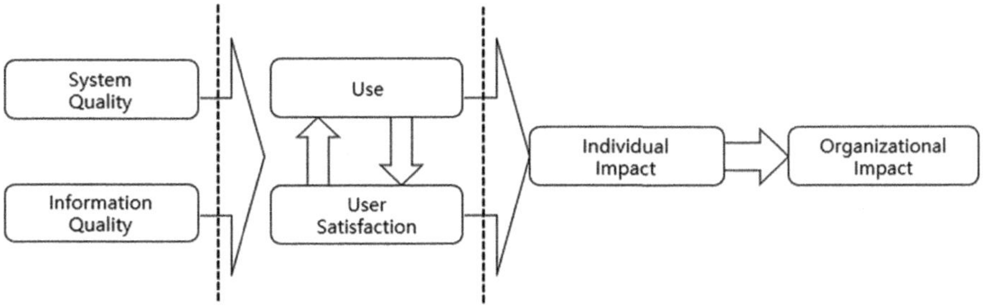

Fig. 1 Delone and Mclean's model of IS success

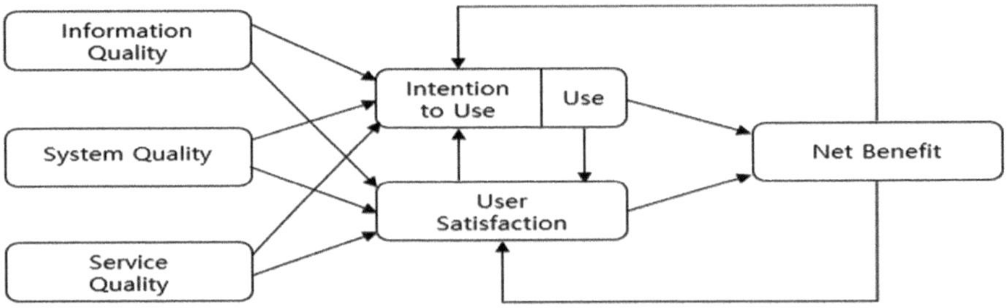

Fig. 2 Updated Delone and Mclean IS success model

As shown in (Fig. 2), the variables of the information system success model are in a mutually influencing relationship. Information quality, system quality, and service quality affect the use and user satisfaction, and use and user satisfaction affect each other. Use and user satisfaction affect net benefits, and net benefit affects use and user satisfaction.

2.4 *Expectation Confirmation*

Continued use of an information system after initial acceptance is a very important factor. Davis et al.'s technology acceptance model (TAM) tried to explain both the acceptance and continuity of information systems in users' technology acceptance behaviour [20]. However, it failed to suggest a solution to the phenomenon of acceptance discontinuance anomaly, in which users stop using the information system again after making a decision to accept the technology in the beginning.

Based on the existing research, an Expectation Confirmation Model was established, and the results were presented that the user satisfaction and perceived usefulness of the information system affect the continuous intention to use [21]. The continuous use of information systems is similar to the decision to repurchase a product. In particular, decisions regarding continued use and repurchase were all made after the initial decision of use and purchase was made, and later through the experience of use, the first decision was made. It was said that the results that contradicted the decision were similar to each other.

TAM aims to reveal the factors that affect the decision to adopt a new information system. In contrast, the Expectation Confirmation Model is a theoretical model to explain the factors that affect the continued use of information systems adopted by consumers. Expectation confirmation model combines expectation disconfirmation theory and TAM to overcome these limitations of TAM. And it is a theory that tries to explain the continuous use of information systems by using expectation satisfaction, perceived usefulness, and satisfaction as independent variables that affect the intention to continue using [22].

Fig. 3 Expectation confirmation model

It was found that satisfaction of expectations confirmation and perceived usefulness had a positive effect on the intention to continue to use through satisfaction. In particular, in the case of expectation confirmation, it was found that it showed greater explanatory power for user satisfaction than perceived usefulness [21]. In the use and satisfaction approach, active users selectively use media that will satisfy their needs among several competing media. And confirmation appears in the form of satisfaction perceived by users as a result of use. In other words, based on the motive of use, the user has expectations for the confirmation to be obtained through use and is satisfied with the acquisition of the confirmation expected as a result of use and decides the intention to use it afterward (Fig. 3).

3 Research Method

3.1 Research Model

In this study, based on the IS Success Model and the expectation confirmation model, NC Dinos' AI, and big data-based dynamic pricing are used to set the spectator factors for professional baseball to identify factors that affect continuous intention to use. The same research model was presented (Fig. 4).

3.2 Hypothesis Settings

3.2.1 Relationship Between Characteristics of Dynamic Price System and Price Fairness

The success factors of information systems are divided into six categories: system quality, information quality, system use, user satisfaction, personal effect, and organizational effect. The quality of information that is used affects the usefulness of the system and user satisfaction [20]. It is evaluated that information system evaluation was carried out from a comprehensive perspective, and it has become the

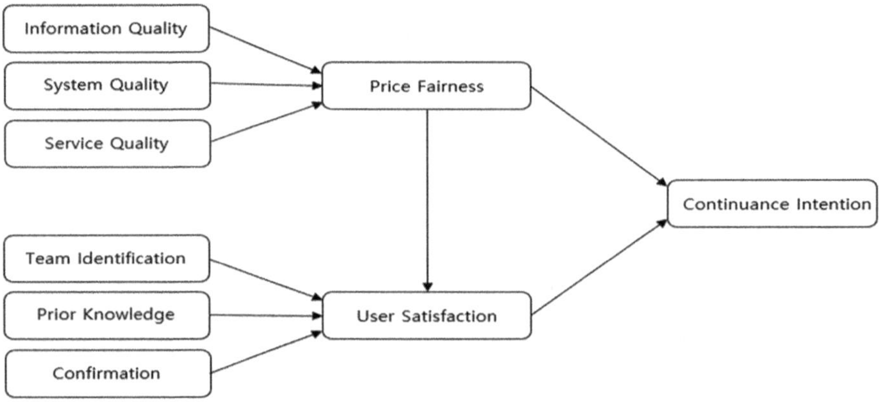

Fig. 4 Research model

basis for many follow-up studies. Because of the close relationship between quality and price fairness, the evaluation of quality can be measured by consumer satisfaction for an appropriate price [23, 24]. Accordingly, this study attempted to improve the completeness of the study by establishing a hypothesis that information quality, system quality, and service quality would affect price fairness based on the study of Delone and Mclean.

H1. Information Quality is positively associated with Price Fairness.
H2. System Quality is positively associated with Price Fairness.
H3. Service Quality is positively associated with Price Fairness.

3.2.2 Relationship Between Attributes of Individual Factors and User Satisfaction

It was said that watching a sports game induces a special psychological response in consumers, and identification occurs in the relationship between sports games and consumers [25]. In addition, since the degree of relationship between oneself and a specific club can have a great influence on consumers' purchasing behavior through individual self-concept, a study on team identification was conducted [26–29]. Consumers can satisfy their desire for self-expression and experience positive emotions through watching sports that are highly congruent with their self-image [30]. In the planned behavior theory, prior knowledge refers to experiences and information related to products that consumers have [31]. Prior knowledge is a very important variable in decision-making, and the lower the prior knowledge, the more difficult it is to process information, and it is said that it is difficult to evaluate the target easily because of lack of experience [32]. The causal relationship that a person with high prior knowledge affects satisfaction was able to confirm the existing research [33–35]. According to Bhattacherjee's study, confirmation has a positive effect on continuance intention through user satisfaction [21]. Therefore, this study tried to increase the completeness of the study by setting a hypothesis that team identification, prior knowledge, and confirmation would affect user satisfaction.

H4. Team Identification is positively associated with User Satisfaction.
H5. Prior Knowledge is positively associated with User Satisfaction.
H6. Confirmation is positively associated with User Satisfaction.

3.2.3 Price Fairness, User Satisfaction and Continuance Intention

It was said that the perception of fairness or unfairness in price affects consumer satisfaction, purchase intention, and complaining behavior [14, 15]. If the price is unfairly perceived by the buyer, the perceived value decreases, and the consumer experiences disappointment [10]. As a result, consumers will complain or try to switch to another seller, and either do not take specific action or decide that it is not worth taking action [36]. When consumers perceive the price of a product or service fairly, positive word of mouth and purchase intention occur [37, 38]. Thus, this study tried to increase the completeness of the study by setting the following hypotheses based on the expectation confirmation model.

H7. Price Fairness is positively associated with User Satisfaction.
H8. Price Fairness is positively associated with Continuance Intention.
H9. User Satisfaction is positively associated with Continuance Intention.

4 Data Analysis

In this study, a total of 128 copies were used to analyze the results by collecting data through a questionnaire targeting the general public who experienced the dynamic pricing of NC Dinos. To test the research model and research hypothesis, frequency analysis and reliability analysis using Cronbach α, DH Rho_A (ρ_A), CR, and intensive validity were evaluated with external loadings and AVE, and discriminant validity was evaluated by Fornell Structural equation analysis was performed with PLS-SEM using the partial least squares method by applying the criteria of the Larcker criterion (Table 1).

4.1 Analysis of Reflective Measurement Model

To test this research model, the internal consistency reliability, convergent validity, and discriminant validity of the factors were confirmed using SMART PLS to confirm whether the questionnaire items were suitable for the purpose and intention of the study. The factor loading, which refers to the degree of correlation between each factor, satisfies the criterion at 0.5 or more.

Table 1 Demographic statistics

Classificant		Frequency	Ratio (%)
Gender	Male	93	72.66
	Female	35	27.34
Age	20's	31	24.22
	30's	43	33.59
	40's	33	25.78
	50's	18	14.06
	Over 60's	3	2.34
Cheering for team	SSG LANDERS	4	3.13
	KIWOOM HEROES	6	4.69
	LG TWINS	13	10.16
	KT WIZ	3	2.34
	KIA TIGERS	13	10.16
	NC DINOS	47	36.72
	SAMSUNG LIONS	9	7.03
	LOTTE GIANTS	21	16.41
	DOOSAN BEARS	7	5.47
	HANWHA EAGLES	5	3.91
Period of cheering number of visits per year	Less than 1 year	3	2.34
	13 years	40	31.25
	4–6 years	30	23.44
	7–9 years	16	12.50
	Over 10 years	39	30.47
	Less than 1 time	5	3.91
	2–5 times	44	34.38
	6–9 times	42	32.81
	Over 10 times	37	28.91
Suppoters status	Yes	45	35.16
	No	83	64.84
Level of education	Less than high school	11	8.59
	College student	9	7.03
	College graduate	89	69.53
	Graduate student	2	1.56
	Graduate school graduate	17	13.28

The Cronbach α value, which evaluates internal consistency reliability, is 0.6 or higher, which meets the recommended criteria [39]. It can be seen that the DH Rho_A (ρ_A) value of 0.7, which is the acceptance criterion, and the CR value of 0.6 or more, are within the high reliability and general acceptance range. In addition, indicator reliability and average variance Extracted (AVE) were found to be more than 0.5, confirming that there is desirable convergent validity [40–46].

The analysis results of internal consistency reliability and convergent validity are shown in (Table 2).

For the criterion for evaluating discriminant validity, the Fornell Lacker criterion, cross-loading, and HTMT are used. Discriminant validity by criteria was considered to have discriminant validity only when the correlation coefficient value between study factors was less than the square root of AVE. According to the Fornell Lacker criterion, each factor compares the square of the Pearson correlation coefficient between the AVE value and the latent variable, and if the AVE value is greater than the square value of the correlation coefficient, it can be judged that the discriminant validity has been secured [44].

The discriminant validity test by HTMT gives an estimate of what the true correlation between factors would be if the two factors were measured perfectly [39] (Tables 3 and 4).

4.2 Analysis of Structural Model

The results of analyzing the model explanatory power for the structural model are shown in (Table 5). The R^2 value is the most common criterion for evaluating the explanatory power of a structural model, and the higher the R^2 value, the higher the prediction accuracy [1]. In this study, all of them can be regarded as having a satisfactory level of explanatory power.

In the evaluation of the structural model, it is necessary to check Stone-Geisser's Q^2 value. This is an evaluation criterion that can determine how predictive fitness is for a specific endogenous variable. If Q^2 value is greater than 0, it is judged that the exogenous latent variable has predictive suitability for the endogenous latent variable as a whole, and if it is less than 0, it is judged to have insufficient predictive suitability [45, 47, 48] (Table 6).

4.3 Hypothesis Testing

The path analysis of the research model can be verified for statistical significance using the t value and P value obtained through Bootstrapping of SMART PLS 4.0. If the t value is greater than the critical values, the corresponding path coefficient is said to be statistically significant. In social science research, $\pm$ 1.96(significance

Table 2 Reliability and validity of latent constructs

Variable	Item	Outer loading	Indicator reliability	AVE	Cronbach α	DH Rho $A(\rho_A)$	CR
Information quality	IQ2	0.800	0.640	0.682	0.767	0.769	0.866
	IQ3	0.844	0.712				
	IQ4	0.833	0.694				
System quality	SyQ1	0.801	0.642	0.716	0.802	0.813	0.883
	SyQ2	0.871	0.789				
	SyQ3	0.865	0.748				
Service quality	SvQ2	0.855	0.731	0.737	0.822	0.824	0.894
	SvQ3	0.848	0.719				
	SvQ4	0.873	0.762				
Team identification	TI1	0.785	0.616	0.675	0.838	0.864	0.892
	TI2	0.730	0.533				
	TI4	0.909	0.826				
	TI5	0.851	0.724				
Prior knowledge	PK1	0.907	0.823	0.779	0.906	0.925	0.934
	PK2	0.889	0.790				
	PK3	0.860	0.740				
	PK4	0.874	0.764				
Confirmation	CF1	0.824	0.679	0.730	0.876	0.879	0.915
	CF2	0.879	0.773				
	CF3	0.857	0.734				
	CF4	0.856	0.733				
Price fairness	PF1	0.901	0.812	0.802	0.918	0.919	0.942
	PF2	0.890	0.792				
	PF3	0.901	0.812				
	PF4	0.892	0.796				
User satisfaction	US1	0.884	0.781	0.790	0.867	0.868	0.919
	US2	0.904	0.817				
	US3	0.879	0.773				
Continuance intention	CI2	0.892	0.796	0.807	0.880	0.881	0.926
	CI3	0.893	0.797				
	CI4	0.910	0.726				

IQ = Information quality, SyQ = System quality, SvQ = Service quality, TI = Team identification, PK = Prior knowledge, CF = Confirmation, PF = Price fairness, US = User satisfaction, CI = Continuance intention

Table 3 Discriminant validity (Fornell-Lacker criterion)

Variable	IQ	SyQ	SvQ	TI	PK	CF	PF	US	CI
IQ	0.826								
SyQ	0.700	0.846							
SvQ	0.687	0.719	0.859						
TI	0.352	0.344	0.327	0.822					
PK	0.382	0.455	0.306	0.501	0.883				
CF	0.545	0.531	0.417	0.480	0.622	0.854			
PF	0.602	0.649	0.632	0.361	0.305	0.481	0.896		
US	0.511	0.483	0.373	0.490	0.482	0.770	0.489	0.889	
CI	0.493	0.416	0.327	0.435	0.450	0.718	0.427	0.784	0.898

Table 4 Discriminant validity (HTMT)

Variable	IQ	SyQ	SvQ	TI	PK	CF	PF	US	CI
IQ									
SyQ	0.884								
SvQ	0.865	0.877							
TI	0.439	0.428	0.390						
PK	0.467	0.541	0.541	0.581					
CF	0.666	0.630	0.630	0.557	0.691				
PF	0.715	0.750	0.750	0.412	0.337	0.538			
US	0.624	0.575	0.575	0.566	0.528	0.883	0.547		
CI	0.604	0.492	0.387	0.498	0.494	0.816	0.475	0.895	

Table 5 Result of R^2 value

Variable	R^2	R^2 adjusted	assessment
Price fairness	0.494	0.482	R^2 0.25 < weak
User satisfaction	0.626	0.614	R^2 0.25 = moderate
Continuance intention	0.616	0.610	R^2 0.25 > strong

Table 6 Result of R^2 value

Variable	SSO	SSE	Q^2 (=1-SSE/SSO)
Price fairness	512	313.342	0.388
User satisfaction	384	200.281	0.478
Continuance intention	384	196.408	0.489

Table 7 Hypothesis assessment

Hypothesis	Path coefficients	T statistics	*p*-value	Label
Information quality → Price fairness	0.192	1.721	0.085	Reject
System quality → Price fairness	0.322	2.837	0.005	**Adopted**
Service quality → Price fairness	0.269	2.552	0.011	**Adopted**
Team identification → User satisfaction	0.145	2.031	0.042	**Adopted**
Prior knowledge → User satisfaction	-0.044	0.545	0.586	Reject
Confirmation → User satisfaction	0.665	8.151	0.000	**Adopted**
Price fairness → User satisfaction	0.129	2.164	0.031	**Adopted**
Price fairness → Continuance intention	0.057	0.909	0.363	Reject
User satisfaction → Continuance intention	0.756	12.040	0.000	**Adopted**

level 5%, $\alpha = 0.05$) is often used, and the research hypothesis is adopted if the p value is smaller than the significance level $\alpha = 0.05$[49] (Table 7).

5 Conclusion

This study tried to understand the mechanism behind the factors of watching professional baseball that affect the continuous intention to use, focusing on NC Dinos to which dynamic pricing is applied.

The goal of this study is to increase and maintain baseball fans by professional baseball teams while exploring the essential parts of professional baseball fans' continuous viewing.

To conduct the study, a questionnaire was used as a measurement tool, and a total of 128 copies were used for analysis. For the survey data, structural equation analysis was performed with PLS-SEM. In order to understand the demographic characteristics of the research subjects, frequency analysis was conducted and each item was factored. The internal consistency reliability, convergent validity, and discriminant validity of the factors were confirmed. The R^2 value, the most common criterion for evaluating the explanatory power of a structural model, was confirmed. The path analysis of the study model was verified for statistical significance using the t value and P value obtained through Bootstrapping of SMART PLS 4.0.

While previous studies mainly focused on identifying the factors that reflected the value of professional baseball, this study is meaningful in that it empirically analyzed the quality of the system in relation to dynamic pricing. Regarding the ticket purchasing system to which dynamic pricing is applied, consumers found that system quality and service quality affect price fairness, but information quality did not. It was not adopted because the audience was not sufficiently aware of the quality.

Personal attributes such as team identification and meeting expectations had an effect on satisfaction, but prior knowledge did not. The higher the level of team identification, the more positive emotions experienced by the team. Thus, it is necessary to develop a strategy to induce positive emotions through connected activities that can increase the sense of unity and belonging to the team. According to a study by Bhattacherjee, confirmation has a positive effect on continuance intention through user satisfaction [37]. Price fairness affects satisfaction, but does not affect continuous intention to use. When considering this comprehensively, it can be inferred that in order for baseball fans to reach viewing behavior, they must have conditions to watch baseball, including user satisfaction, rather than other attributes. In addition, since watching baseball forms a positive attitude, it is necessary to ensure that price fairness conditions and satisfaction using dynamic pricing lead to viewing behavior.

This study has limitations in generalization as it only targets professional baseball among spectator sports. Therefore, if more diverse sports events are conducted, it will be of great help in understanding the behavior of spectators.

In addition, a study was conducted based on visitors who visited the home stadium of the NC Dinos and experienced dynamic pricing. In future studies, it is expected that a more complete study will be possible if a study based on various demographic characteristics is conducted while studying each club or supporters and non-supporters on the same line.

References

1. Hill B, Green BC (2000) Repeat attendance as a function of involvement, loyalty, and the sportscape across three football contexts. Sport Manag Rev 3(2):145–162
2. McGill JI, Van Ryzin CI (1999) Revenue management: research overview and prospects. Transp Sci 33(2):233–256
3. Paul RJ, Weinbach AP (2013) Determinants of dynamic pricing premiums in major league baseball. Sport Mark Q 22(3):152–165
4. Barilla AG, Gruben K, Levernier W (2008) The effect of promotions on attendance at major league baseball games. J Appl Bus Res (JABR) 24(3):1–14
5. Lemke RJ, Leonard M, Tlhokwane K (2010) Estimating attendance at major league baseball games for the 2007 season. J Sports Econ 11(3):316–348
6. Kannan PK, Praveen KK (2001) Dynamic pricing on the internet: importance and implications for consumer behavior. Int J Electron Commer 5(3):63–68
7. Marshall M (2001) Dynamic pricing for e-commerce, an integrated solution. In: IEEE In Proceedings Third International Workshop on Advanced Issues of E-Commerce and Web-Based Information Systems, pp 192–194
8. Proshansky HM, Ittelson WH, Rivlin LG (1972) Freedom of choice and behavior in a physical setting
9. Adams JS (1963) Toward an understanding of inequity. J Abnormal Soc Psychol 67(5):422–436
10. Xia L, Monroe KB, Cox JL (2004) The price is unfair! a conceptual framework of price fairness perceptions. J Mark 68(4):1–15
11. Bolton LE, Warlop L, Alba JW (2003) Consumer perceptions of price (un) fairness. J Cons Res 29(4):474–491
12. Kahneman D, Knetsch JL, Thaler R (1986) Fairness as a constraint on profit seeking: entitlements in the market. The Am Econ Rev, 728–741

13. Oliver RL, Swan JE (1989) Consumer perceptions of interpersonal equity and satisfaction in transactions: a field survey approach. J Mark 53(2):21–35
14. Campbell MC (1999) Perceptions of price unfairness: antecedents and consequences. J Mark Res 36(2):187–199
15. Huppertz JW, Arenson SJ, Evans RH (1978) An application of equity theory to buyer-seller exchange situations. J Market Res 15(2):250–260
16. DeLone WH, McLean ER (1992) Information systems success: the quest for the dependent variable. Information Syst Res 3(1):60–95
17. Ballantine J, Bonner M, Levy M, Martin A, Munro I, Powell PL (1996) The 3-D model of information systems success: the search for the dependent variable continues. Inform Res Manage J (IRMJ) 9(4):5–15
18. Pitt LF, Watson RT, Kavan CB (1995) Service quality: a measure of information systems effectiveness. MIS Quart, 173–187
19. DeLone WH, McLean ER (2003) The DeLone and McLean model of information systems success: a ten-year update. J Manag Inf Syst 19(4):9–30
20. Davis FD, Bagozzi RP, Warshaw PR (1989) User acceptance of computer technology: a comparison of two theoretical models. Manag Sci 35(8):982–1003
21. Bhattacherjee (2001) Understanding information systems continuance: an expectation-confirmation model. MIS Quart 25(3):351–370
22. Oliver RL (1989) A cognitive model of the antecedents and consequences of satisfaction decisions. J Mark Res 17(4):460–469
23. Dodds WB, Monroe KB, Grewal D (1991) Effects of price, brand, and store information on buyers' product evaluations. J Mark Res 28(3):307–319
24. Zeithaml VA, Berry LL, Parasuraman LA (1988) Communication and control processes in the delivery of service quality. J Market 52(2):35–48
25. Real MR, Mechikoff RA (1992) Deep fan: Mythic identification, technology and advertising in spectator sports. Sociol Sport J 9(4):323–339
26. Bizman A, Yinon Y (2002) Engaging in distancing tactics among sport fans: effects on self-esteem and emotional responses. J Soc Psychol 142(3):381–392
27. Fink JS, Trail GT, Anderson DF (2002) Environmental factors associated with spectator attendance and sport consumption behavior: gender and team differences. Sport Mark Q 11(1):8–19
28. Trail GT, Fink JS, Anderson DF (2003) Sport spectator consumption behavior. Sport Mark Q 12(1):8–17
29. Wann DL, Royalty JL, Rochelle AR (2002) Using motivation and team identification to predict sport fans' emotional responses to team performance. J Sport Behav 25(2):207–216
30. Swann WB, De La Ronde C, Hixon JG (1994) Authenticity and positivity strivings in marriage and courtship. J Pers Soc Psychol 66(5):857–869
31. Duhan DF, Johnson SD, Wilcox JB, Harrell GD (1997) Influences on consumer use of word-of-mouth recommendation sources. J Acad Mark Sci 25(4):283–295
32. Hoch SJ, Deighton (1989) Managing what consumers learn from experience. J Market 53(2):1–20
33. Leonard D, Sensiper S (1998) The role of tacit knowledge in group innovation. Calif Manage Rev 40(3):112–132
34. Liebeskind JP (1996) Knowledge, strategy, and the theory of the firm. Strateg Manag J 17(S2):93–107
35. Tonglet M, Phillips PS, Read AD (2004) Using the theory of planned behaviour to investigate the determinants of recycling behaviour: a case study from Brixworth, UK. Resour Conserv Recycl 41(3):191–214
36. Urbany JE, Madden TJ, Dickson PR (1989) All's not fair in pricing: an initial look at the dual entitlement principle. Mark Lett 1:17–25
37. Bei LT, Chiao YC (2001) An integrated model for the effects of perceived product, perceived service quality, and perceived price fairness on consumer satisfaction and loyalty. J Cons Satisfac, Dissatisfac Complain Behav 14:125–140

38. Bujisic M, Bilgihan A, Hutchinson J (2004) The effect of the type-pricing strategy on perceived price fairness and behavioral outcomes in beverage establishments. J Revenue Pricing Manage 13:35–60
39. Hair JF, Hult GTM, Ringle CM, Sarstedt M (2021) A primer on partial least squares structural equation modeling (PLS-SEM). Sage Publications
40. Buzeta C, De Pelsmacker P, Dens N (2020) Motivations to use different social media types and their impact on consumers' online brand-related activities (COBRAs). J Interact Mark 52(1):79–98
41. Chin WW (1998) The partial least squares approach to structural equation modeling. Modern Methods Bus Res 295(2):295–336
42. Cronbach LJ (1951) Coefficient alpha and the internal structure of tests. Psychometrika 16(3):297–334
43. Dijkstra TK, Henseler J (2015) Consistent and asymptotically normal PLS estimators for linear structural equations. Comput Stat Data Anal 81:10–23
44. Fornell C, Larcker DF (1981) Evaluating structural equation models with unobservable variables and measurement error. J Mark Res 18(1):39–50
45. Geisser S (1974) A predictive approach to the random effect model. Biometrika 61(1):101–107
46. Nunally JC, Bernstein IH (1994) Psychometric theory. McGraw-Hill, New York
47. Cha J (1994) Partial least squares. Adv Methods Mark Res 407:52–78
48. Stone M (1997) An asymptotic equivalence of choice of model by cross-validation and Akaike's criterion. J Roy Stat Soc: Ser B (Methodol) 39(1):44–47
49. Efron B, Tibshirani R (1997) Improvements on cross-validation: the 632+ bootstrap method. J Am Stat Assoc 92(438):548–560

Author Index

R. Lee et al. (eds.), *AI and Metaverse*, Studies in Computational Intelligence 1160,
https://doi.org/10.1007/978-3-031-61163-6

Printed by Printforce, the Netherlands